GUO Wei 郭炜 著

NEW GENERATION OF REGIONAL AIRPORT TERMINAL BUILDINGS
ARCHITECTURAL DESIGN AND INNOVATIVE PATHS

新一代支线机场航站楼
建筑设计与创新路径

同济大学出版社·上海
Tongji University Press · Shanghai

图书在版编目（CIP）数据

新一代支线机场航站楼：建筑设计与创新路径 = New Generation of Regional Airport Terminal Buildings: Architectural Design and Innovative Paths：英文 / 郭炜著 . -- 上海：同济大学出版社，2024. 11. -- ISBN 978-7-5765-1283-0

I. TU248.6

中国国家版本馆 CIP 数据核字第 2024AJ0939 号

New Generation of Regional Airport Terminal Buildings: Architectural Design and Innovative Paths

新一代支线机场航站楼：建筑设计与创新路径

郭 炜 著

策划编辑	胡 毅 晁 艳
责任编辑	王胤瑜
责任校对	徐逢乔
装帧设计	完 颖
排　　版	朱丹天

出版发行　同济大学出版社　www.tongjipress.com.cn
　　　　　（地址：上海市四平路1239号　邮编：200092　电话：021-65985622）
经　　销　全国各地新华书店、网络书店
印　　刷　上海安枫印务有限公司
开　　本　889mm × 1194mm　1/20
印　　张　17
字　　数　380 000
版　　次　2024 年 11 月第 1 版
印　　次　2024 年 11 月第 1 次印刷
书　　号　ISBN 978-7-5765-1283-0
定　　价　208.00 元

本书若有印装质量问题，请向本社发行部调换　　　版权所有　　侵权必究

Foreword

Airport is an important infrastructure for human society. It is a carrier for the speedy exchange and flow of people, logistics and information flow.

With the rapid development of China's aviation industry, a large number of excellent airport terminal buildings have emerged in recent years. However, we should not only pay attention to the large international airports, but also realize the significance of regional airports which account for a considerable proportion of the airports in China, and should make great efforts to develop regional airports to build a modern, convenient and dense network of regional airports. This helps lay a solid foundation for China's transition from a single air transportation power to a multi-disciplinary civil aviation power, and accelerate the shift to high-quality development of civil aviation. It requires breakthroughs in the planning and design theory of small- and medium-sized airport terminal buildings.

At present, research related to terminals of small- and medium-sized regional airports is of extreme scarcity. As a response, this book focuses on architectural design of small- and medium-sized regional airport terminals, systematically analyzes the existing problems of regional airport design, points out the necessity of innovation, and sorts out the development process of China's regional airports into four "generations," putting forward the "fourth generation" of terminal buildings that have gone beyond the "era of function," "era of style," and "era of culture" to enter the "era of experience." On analyzing the characteristics and deficiencies of the previous three generations, the author puts forward the concept of the fourth-generation "experiential" regional airport terminal. In addition, the author has studied and summarized the excellent regional airport terminal cases in China and many other countries and regions, providing reference for the future construction of China's regional airports.

Given that most existing research on regional airports in China has not yet touched on the design methods of terminal building in the new era, this book explores the design concepts and methods to create excellent experience in terminals based on practice of the author's team, and proposes an evaluation system on innovativeness of regional airport terminal design in the new era, filling the research gap in the field of small airport terminal design for China.

This book systematically presents key outlines in design of small- and medium-sized regional airport terminals, summarizes the general rules and concepts in design, which provides guidance for improving the architectural grace, artistic value, and cultural experience of small- and medium-sized regional airport terminal buildings.

Being rare both at home and abroad, the research in this book will play a leading role in improving the design level of small- and medium-sized airport terminals in China.

National Design Master of Geotechnical Investigation and Survey
General Architect of ECADI

Preface

Accompanied by the rapid economic development of Reform and Opening Up for more than 40 years, China's aviation industry is rising rapidly. As of the beginning of 2022, China's licensed transport airports have reached 248 (excluding Hong Kong, Macao, and Taiwan), the share of passenger throughput of airports in the central and western parts of the country has risen to 44.4%, with a more strategic position, and the civil aviation passenger turnover in the comprehensive transportation has risen to 33%, and a rational layout has been built basically with perfect function, safety and high efficiency.

The "Plan of National Civil Transportation Airport Layout" proposes to continue to promote the natioanwide layout of airports, and by 2025, 136 new airports will be added, with a total of 370 civil transportation airports planned for across the country (about 320 airports are planned to be built).

The "'14th Five-Year Plan' for Civil Aviation Development" states that during the "14th Five-Year Plan" period, China's civil aviation development will be anchored on becoming a civil aviation power, preparing for the leap from a single civil aviation transportation power to a multi-disciplinary civil aviation power. This requires China's civil aviation to accelerate its transition to high-quality development, which also means that the aviation market still has huge growth potential. Overall, the fundamentals supporting the sustained and rapid growth of China's civil aviation during the "14th Five-Year Plan" period have not changed, and the main contradiction between the unbalanced and insufficient development of civil aviation and people's ever-increasing demand for air travel has not changed, either. The development of China's civil aviation industry is still facing strategic opportunity.

The plan also clearly points out that during the "14th Five-Year Plan" period, it is necessary to improve the layout of non-hub airports, building a number of new non-hub airports to intensifiy the network layout. Specifically, the 23 new airports planned for the "14th Five-Year Plan" period are all regional airports, and 57 regional airports are upon new construction and relocation preliminary work. It can be said that in the civil aviation "14th Five-Year Plan" period, the construction of regional airports is the core task. At present, the design, construction and management of gateway airports and regional hub airports of China has been in line with international standards, but those of regional airports still need to be improved.

As a pioneer in airport design and construction in China, East China Architectural Design & Research Institute Co., Ltd. (ECADI) participated in the construction of Shanghai Hongqiao International Airport in the 1960s. After Reform and Opening Up, it started from the first phase of Shanghai Pudong International Airport project, and continued to work in the field of airport construction for more than 30 years, successively participating in the second phase of Shanghai Pudong International Airport, and the satellite hall of Shanghai Pudong International Airport, Hohhot Airport in Inner Mongolia, Taiyuan Airport in Shanxi Province, Hefei Airport and Bengbu Airport in Anhui Province, Kunming Airport, Dali Airport, and Lancang Airports in Yunnan Province, North Terminal Area of Urumqi Airport in Xinjiang, Dingri Airport in Xizang, etc., accumulating rich experience in engineering design for airport construction. The high level of theoretical precipitation has laid a solid foundation for the writing of this book.

From the perspective of terminal building design, this book combs through the relevant standards and technical points of regional airport design in China, covering almost all the technical details in the planning and design of small regional airports. At the same time, it summarizes the excellent airport terminal design cases both in China and abroad in recent years, and specifically analyzes their design concepts and techniques, which can provide abundant references for the construction of future regional airports. Meanwhile, this book explores the design concepts, ideas and methods of excellent airport terminal buildings in the new era combined with the airport design practice of the author's team, proposing the evaluation aspects of innovativeness of the design of regional airport terminal buildings, making contribution to innovative design of China's regional airport terminal buildings.

This book is a thematic study on small regional airports after the benchmarking project "Comprehensive Technical Study of Airports" of Arcplus Group PLC (hereafter referred as Arcplus), also as a supplement to that project by further enriching study on airports and improving the related technical reserve. The publication of this book will show the professionalism and competitiveness of Arcplus in the market segment, and bring reference for the front-line designers.

The writing of this book was supported by: Guo Jianxiang, National Design Master of Geotechnical Investigation and Survey, and General Architect of ECADI; Wang Dasui, National Design Master of Geotechnical Investigation and Survey, and Senior General Engineer of ECADI; Shen Di, National Design Master of Geotechnical Investigation and Survey, and Chief General Architect of Arcplus; Zhang Junjie, Vice Managing Director of the Architectural Society of Shanghai China, and General Architect of ECADI; Liu Wujun, former General Engineer of Shanghai Airports (Group) Co., Ltd.; Zhu Jingyuan, former Deputy General Director of Construction Command of Capital Airport Group Co., Ltd.; and Gao Wenyan, General Architect of Shanghai Scientific Technology Development Branch of Arcplus. For their advice and guidance, the author's team would like to express sincere respect and deep gratitude!

CONTENTS

FOREWORD
PREFACE

CONTENTS

1 INTRODUCTION 1

1.1 Scope and Definition of Regional Airport 3
1.2 Fundamental Study of One-and-a-Half-Story Terminal Building of Regional Airport 18
1.3 Significance of Design Research on Regional Airport Terminal Building 21

2 PLANNING
General Composition and Basic Elements of Regional Airport 23

2.1 Spatial Composition of Airport 24
2.2 Basic Introduction to Airfield Area 25
2.3 Basic Introduction to Terminal Area 31

3 FUNCTIONS
Functional Composition and Design Outlines of Terminal Building 51

3.1 Functional Areas of Terminal and Passenger Flows 52
3.2 Design Outlines of Arrival/Departure Area 66
3.3 Design Outlines of Check-in and Ticketing Area 71
3.4 Design Outlines of Security Checkpoint/Joint Inspection Area 82
3.5 Design Outlines of Boarding Lounge 93
3.6 Design Outlines of Boarding Gate 96
3.7 Design Outlines of Baggage Claim Hall 100
3.8 Design Outlines of Baggage Handling Area 105
3.9 Design Outlines of VIP Lounge 113

4 SPATIAL COMPOSITION
Spatial Composition of One-and-a-Half-Story Terminal Building 121

4.1 Types of Spatial Combination of One-and-a-Half-Story Terminal Building 122
4.2 Spatial Scale Study of One-and-a-Half-Story Terminal 129

5 INNOVATION
Design Innovation of Regional Airport Terminal Building 151

5.1 Problems in Design of Terminal Building 152
5.2 Innovative Methods and Excellent Cases of Terminal Building Design 153

6 TENDENCY
Four Generations of Regional Airport Terminal and Innovative Practices of the Fourth Generation (Era of Experience) 215

6.1 Four Generations of Regional Airport Terminal Design 216
6.2 Innovative Design Practices of the Fourth-Generation Regional Airport Terminal 217

7 SUMMARY
Innovative Concepts and Evaluation of the Fourth-Generation (Era of Experience) Regional Airport Terminal 291

7.1 Concepts and Methods for Innovative Design of the Fourth-Generation Regional Airport Terminal 292
7.2 Evaluation Aspects and Methods for Design of the Fourth-Generation Regional Airport Terminal 296

ILLUSTRATION SOURCES 319
POSTSCRIPT 323

1 INTRODUCTION

As China's aviation industry has experienced rapid growth, the number of civil airports has surged significantly. As of December 31, 2021, China had certified 248 transportation airports, excluding those in Hong Kong, Macau, and Taiwan. These airports cover 91.7% of the country's prefecture-level cities, and the civil aviation fleet of the whole nation has expanded to 6,795 aircraft.

In recent years, there have been continuous improvements in quality and efficiency, as well as the increasing importance of Chinese civil aviation transportation. The passenger throughput percentage of airports in central and western China has increased to 44.4% of the national total, and its share of passenger volume in the comprehensive transportation network has risen to 33%. Meanwhile, there are now 895 international routes reaching 62 countries, indicating that a well-structured, functional, safe, and efficient airport network has been established in China[1]. During the "13th Five-Year Plan" period, China essentially achieved a historic transition from being a major transportation country to becoming a transportation powerhouse.

Furthermore, based on "Plan of National Civil Transportation Airport Layout", by 2025, China plans to add 136 new airports, bringing the total number of planned civil transport airports to 370 nationwide, with about 320 of them expected to be completed then.

The "'14th Five-Year Plan' for Civil Aviation Development" highlights that the development of the civil aviation industry shall anchor around the strategic objective of transforming China into a multifaceted leading power in civil aviation industry, which demands an accelerated transformation towards high-quality development. This implies significant growth potential and crucial strategic opportunities for China's civil aviation market.

Overall, during the "14th Five-Year Plan" period, the fundamental mission to continuously support the rapid growth of China's civil air transportation has not changed. The main contradiction between the unbalanced and insufficient provision of civil air transportation services and the continuously growing needs for air travel of the Chinese people also remains unchanged. According to the "14th Five-Year Plan," during this period, the layout of non-hub airports in China will be improved, which means a number of new non-hub airports will be built, with a focus on increasing the density of those in the central and western regions and border areas. Notably, the 23 new airports planned for the "14th Five-Year Plan" period are all regional airports; 3 of the 4 airports planned for relocation are regional airports; and the preliminary work on the new construction and relocation of another 57 regional airports will launch. It can

---

[1] Data sources: *National Civil Transportation Airport Production Statistics Bulletin 2021 of China*, and the "'14th Five-Year Plan' for Civil Aviation Development."

be said that during the "14th Five-Year Plan" period, the construction of regional airports is the main focus of China's civil aviation development.

1.1 Scope and Definition of Regional Airport

1.1.1 Classification of China's Airport Systems

From the perspective of national strategies, the "Plan of the National Civil Transportation Airport Layout" mainly considers the overall economic and social development, the integration of aviation transportation with various modes of transportation, and other related factors. It categorizes transportion airports into five levels: world-class airport clusters, international hub airports, regional hub airports, mainline airports, and regional airports. However, in specific practices, various sectors interpret airport categorization differently based on their specific focus areas.

In the realm of operation and management, the Ministry of Finance of the People's Republic of China and the Civil Aviation Administration of China issued a notice on April 30, 2020, titled "Notification on the Amendment of the Interim Measures for the Management of Subsidies at Small and Medium Civil Aviation Airports." This notice adjusted the subsidy scope for China's small and medium civil airports from those "with an annual passenger throughput of under 3 million person-times" to those "under 2 million person-times." Additionally, this revised document categorized China's airports into three regions based on location (eastern, central, or western region) and six standards based on annual passenger throughput (1.5 million ~ 2 million, 1 million ~ 1.5 million, 0.5 million ~ 1 million, 0.3 million ~ 0.5 million, 0.1 million ~ 0.3 million, and no more than 0.1 million person-times), totaling 18 subcategories for subsidies. Therefore, airports with passenger throughput of less than 2 million person-times can be identified as "small- to medium-sized." This definition also aligns with the scope of airports eligible for feeder airline subsidies described in the "Notification on Issuing Amended Interim Management Measures of Subsidies for Feeder Airline Services by Civil Aviation Administration of China" (Civil Aviation Document [2013] No. 28).

From the investment management perspective, the "Interim Measures for the Management of Investment Subsidies for Civil Aviation Infrastructure Projects" categorizes civil airports based on annual passenger throughput into five types: over 50 million, 9.5 million to 50 million (including), 1.8 million to 9.5 million (including), 500,000 to 1.8 million (including), and less than 500,000 person-times.

Focusing on construction management, the "Construction Standards of Civil Airport Engineering Project" issued in 2008 classifies civil airports by passenger throughput into seven levels. It specifies that Levels 1 and 2 (with annual passenger throughput less than 500,000 person-times) are applicable to the

"Construction Standards of Civil Aviation Regional Airports" (Table 1-1). In this context, airports with an annual throughput under 500,000 person-times can be identified as "regional airports," primarily serving short-haul flights, with planned direct flight distances generally ranging from 800 to 1,500 kilometers.

Table 1-1   Construction Standards for Terminal Area of China's Airports of Different Scales

Index	Annual Passenger Throughput P (in ten thousand person-times)	Applicable Standards
1	$P < 10$	"Construction Standards of Civil Aviation Regional Airports"
2	$10 \leqslant P \leqslant 50$	
3	$50 < P < 200$	"Civil Airport Construction Project Standards"
4	$200 \leqslant P < 1000$	
5	$1000 \leqslant P < 2000$	
6	$2000 \leqslant P < 4000$	
7	$4000 \leqslant P$	Individual approval

Note: Data is sourced from "Civil Airport Construction Project Standards."

In the realm of planning management, "The '13th Five-Year Plan' for the Development of Civil Aviation in China (2016—2020)" classifies China's transportation airports into four types based on the national strategies of "The Belt and Road" initiative, the coordinated development of Beijing-Tianjin-Hebei, and the Yangtze River Economic Belt. These types are large/medium/small hub airports, and non-hub airports. The specific definitions are outlined in Appendix 2 of "The '13th Five-Year Plan' for the Construction of Civil Transportation Airports," titled "National Comprehensive Airport System Classification Framework," in which airports are categorized based on their type and the proportion of annual passenger volume to the national total transportation volume:

① Airports with passenger throughput exceeding 1% of national total passenger throughput are classified as large hub airports, with international passenger throughput accounting for more than 5% of national total international passenger throughput;

② Airports with passenger throughput exceeding 1% of national total passenger throughput are classified as medium hub airports;

③ Airports with passenger throughput ranging from 0.2% to 1% of national total passenger throughput are classified as small hub airports;

④ Airports with passenger throughput less than 0.2% of national total passenger throughput are classified as non-hub airports.

Notably, this classification standard does not include distinctions between mainline airports and feeder airports. (Table 1-2)

Table 1-2 National Comprehensive Airport System Classification Framework of China

Transport Airport Categories	Classification Standards	Functional Attributes	Descriptions
Large hub airports	The annual passenger throughput of which accounts for more than 1% of the national total transport volume, and international passenger throughput of which accounts for more than 5% of the national total international passenger throughput	International hub airports	The cumulative international passenger throughput of large hub airports accounts for more than 60% of the national total transport volume
Medium hub airports	The annual passenger throughput of which accounts for more than 1% of the national total transport volume	Regional hub airports	The cumulative passenger throughput of large and medium hub airports accounts for more than 80% of the national total transport volume
Small hub airports	The annual passenger throughput of which accounts for 0.2% to 1% of the national total transport volume	Local hub airports	The cumulative passenger throughput of large, medium and small hub airports accounts for more than 95% of the national total transport volume
Non-hub airports	The annual passenger throughput of which accounts for less than 0.2% of the national total transport volume	Non-hub airports	The cumulative passenger throughput of non-hub airports accounts for no more than 5% of the national total transport volume

Note: this table is sourced from Appendix 2 of "The '13th Five-Year Plan' for the Construction of Civil Transportation Airports."

In terms of the construction grade of the airfield area, according to the latest "Technical Standards on Civil Airport Airfield Area," airport airfield areas are classified by Criteria I and II based on the characteristics of aircraft intended to use the airfield area. Criterion I distinguishes Categories 1, 2, 3, and 4 based on the longest reference length required by all types of aircraft using the runways of that airfield area. Criterion II distinguishes Categories A, B, C, D, E, and F based on the maximum wingspan, or the maximum spacing between the outer edges of the outer wheels of the main landing gear of all types of aircraft using the runways. The higher requirement between the two determines the grade of the airfield area, as indicated in Table 1-3.

Table 1-3 Airfield area Criteria I and II in "Technical Standards on Civil Airport Airfield Area" (in meters)

Airfield Area Criterion I	
Category	Aircraft Reference Airfield Length
1	< 800
2	800 ~ 1200 (excluded)
3	1200 ~ 1800 (excluded)
4	≥ 1800

Airfield Area Criterion II		
Category	Maximum Wingspan of Aircraft	Maximum Spacing between the Outer Edges of the Outer Wheels of the Main Landing Gear
A	< 15	<4.5
B	15 ~ 24 (excluded)	4.5 ~ 6 (excluded)
C	24 ~ 36 (excluded)	6 ~ 9 (excluded)
D	36 ~ 52 (excluded)	9 ~ 14 (excluded)
E	52 ~ 65 (excluded)	9 ~ 14 (excluded)
F	65 ~ 80 (excluded)	14 ~ 16 (excluded)

Currently, the majority of airports in China are classified as 4F, 4E, 4D, and 4C. Considering the corresponding aircraft types for each airfield area grade, 4F airports can accommodate all current aircraft types; 4E airports can serve aircraft such as A330, B747, and smaller types ; 4D airports can serve A300, B767, and smaller types; and 4C airports can serve A320, B737, and smaller types.

1.1.2 Definition of Regional Airports

As described above, the definitions of regional airports in China vary according to different standards, but they are mainly based on factors such as annual passenger throughput, types of aircraft accommodated, and the location of airport. Considering timeliness of the standards and their guidance value on actual design, this book categorizes airports based on the Appendix 2 of "The '13th Five-Year Plan' for the Construction of Civil Transportation Airports," "National Comprehensive Airport System Classification Framework," with additional indicators such as actual annual passenger throughput, terminal area, and airfield area grade. This categorization is derived from the investigation of existing built-up transport airports, aiming to provide a clearer and more comprehensive classification system of airports with an explicit definition for regional airports.

By reviewing the data on the passenger throughput of national civil transport airports in recent years[1], it is possible to estimate the passenger throughput range, terminal scale range, and airfield area grade of these airports. Taking data from 2019 to 2021, this book classifies China's airports according to the indicators proposed above, and lists the basic information of representative airports of each category for the three years, respectively[2].

In 2019, the total passenger throughput of all civil airports in China was 1,351,628,545. Based on the proposed classification standards, the airports are categorized as follows.

12 large hub airports: Located in Beijing, Shanghai, Guangzhou, Chengdu, Shenzhen, Kunming, Xi'an, Chongqing, Urumqi, Harbin, etc., with annual passenger throughput ranging from 20 million to 100 million person-times (excluding the newly opened Beijing Daxing International Airport), terminal building area ranging from 244,000 to 1.41 million square meters, and airfield area grades of 4F and 4E.

---

[1] Data source: Civil Aviation Administration of China, http://www.caac.gov.cn/so/s?siteCode=bm70000001&tab=xxgk&qt=2020%E5%B9%B4%E6%9C%BA%E5%9C%BA%E5%90%9E%E5%90%90%E9%87%8F%E6%8E%92%E5%90%8D

[2] Data in the follwing Tables 1–4 ~ 1–6 is retrieved from the Annual Airport Production Bulletins released by the Civil Aviation Administration of China from 2019 to 2021, respectively, and various airport official websites.

20 regional hub airports: Located in Hangzhou, Shenyang, Jinan, Lanzhou, Nanning, Guiyang, Lhasa, Nanchang, etc., with annual passenger throughput ranging from 13.5 million to 40 million person-times, terminal building area ranging from 81,000 to 614,000 square meters, and airfield area grades of 4F and 4E.

26 small hub airports: Located in Hohhot, Ordos, Xishuangbanna, Zhangjiajie, etc., with annual passenger throughput ranging from 2.7 million to 13.5 million person-times, terminal building area ranging from 14,000 to 209,000 square meters, and airfield area grades of 4E, 4D, and 4C.

179 non-hub airports: Located in Dehong, Enshi, etc., with annual passenger throughput below 2.7 million person-times, terminal building area ranging from 500 to 105,300 square meters, and airfield area grades of 4E, 4D, 4C, and 3C. (Table 1-4)

Table 1-4 Civil Airports Classification of China in 2019

Category	Ranking	Representative Airport	Passenger Throughput (person-times)	Passenger Throughput Range (person-times)	Terminal Building Area Range (m^2)	Airfield Area Grade
Large hub airports (12)	1	Beijing Capital International Airport	100,013,642	20 million ~ 100 million (taking international passengers into consideration)	244,000 ~ 1.41 million	4F, 4E
	2	Shanghai Pudong International Airport	76,153,455			
	3	Guangzhou Baiyun International Airport	73,378,475			
	4	Chengdu Shuangliu International Airport	55,858,552			
	5	Shenzhen Bao'an International Airport	52,931,925			
	6	Kunming Changshui International Airport	48,075,978			
	7	Xi'an Xianyang International Airport	47,220,547			

continued

Category	Ranking	Representative Airport	Passenger Throughput (person-times)	Passenger Throughput Range (person-times)	Terminal Building Area Range (m²)	Airfield Area Grade
Large hub airports (12)	8	Shanghai Hongqiao International Airport	45,637,882	20 million ~ 100 million (taking international passengers into consideration)	244,000 ~ 1.41 million	4F, 4E
	9	Chongqing Jiangbei International Airport	44,786,722			
	18	Urumqi Diwopu International Airport	23,963,167			
	21	Harbin Taiping International Airport	20,779,745			
	53	Beijing Daxing International Airport	3,135,074①	/		
Regional hub airports (20)	10	Hangzhou Xiaoshan International Airport	40,108,405	13.5 million ~ 40 million	81,000 ~ 614,000	4F, 4E
	31	Nanchang Changbei International Airport	13,637,151			
Small hub airports (26)	32	Hohhot Baita International Airport	13,151,840	2.7 million ~ 13.5 million	14,000 ~ 209,000	4E, 4D, and 4C
	59	Ordos Ejin Horo International Airport	2,695,925			
Non-hub airports (179)	60	Beihai Fucheng Airport	2,679,101	Below 2.7 million	500 ~ 105,300	4E, 4D, 4C, and 3C
	238	Changhai Dachangshandao Airport	3,260			

Note: ① There was lower passenger throughput of Beijing Daxing International Airport in 2019 because it just was opened that year.

In 2020, due to the impact of the COVID-19 pandemic, Chinese aviation passenger traffic decreased by 36.6% compared to 2019, and the total passenger throughput nationwide was 857,159,437 person-times. Based on the proposed classification standard, the airports are categorized as follows.

12 large hub airports: Located in Beijing, Shanghai, Guangzhou, Chengdu, Shenzhen, Kunming, Xi'an, Chongqing, Urumqi and Harbin, with annual passenger throughput ranging from 8.5 million to 40 million person-times, terminal building area ranging from 244,000 to 1,410,000 square meters, and airfield area grades of 4F and 4E.

23 regional hub airports: Located in Hangzhou, Shenyang, Jinan, Lanzhou, Nanning, Guiyang, Lhasa, Dalian,etc., with annual passenger throughput ranging from 8.5 million to 28 million person-times, terminal building area ranging from 81,000 to 614,000 square meters, and airfield area grades of 4F and 4E.

24 small hub airports: Located in Shijiazhuang, Hohhot, Xishuangbanna, Weihai,etc., with annual passenger throughput ranging from 1.7 million to 8.5 million person-times, terminal buidling area ranging from 13,000 to 209,000 square meters, and airfield area grades of 4E, 4D, and 4C.

181 non-hub airports: located in Yancheng, Dehong, Enshi, etc., with annual passenger throughput less than 1.7 million person-times, terminal building area ranging from 500 to 55,700 square meters, and airfield area grades of 4E, 4D, 4C, and 3C. (Table 1-5)

Table 1-5 Civil Airports Classification of China in 2020

Category	Ranking	Representative Airport	Passenger Throughput (person-times)	Passenger Throughput Range (person-times)	Terminal Building Area Range (m^2)	Airfield Area Grade
Large hub airports (12)	1	Guangzhou Baiyun International Airport	43,760,427	8.5 million ~ 40 million (taking international passengers into consideration)	244,000 ~ 1.41 million	4F, 4E
	2	Chengdu Shuangliu International Airport	40,741,509			
	3	Shenzhen Bao'an International Airport	37,916,059			
	4	Chongqing Jiangbei International Airport	34,937,789			

continued

Category	Ranking	Representative Airport	Passenger Throughput (person-times)	Passenger Throughput Range (person-times)	Terminal Building Area Range (m²)	Airfield Area Grade
Large hub airports (12)	5	Beijing Capital International Airport	34,513,827	8.5 million ~ 40 million (taking international passengers into consideration)	244,000 ~ 1.41 million	4F, 4E
	6	Kunming Changshui International Airport	32,989,127			
	7	Shanghai Hongqiao International Airport	31,165,641			
	8	Xi'an Xianyang International Airport	31,073,884			
	9	Shanghai Pudong International Airport	30,476,531			
	17	Beijing Daxing International Airport	16,091,449			
	20	Harbin Taiping International Airport	13,508,687			
	25	Urumqi Diwopu International Airport	11,152,723			
Regional hub airports (23)	10	Hangzhou Xiaoshan International Airport	28,224,342	8.5 million ~ 28 million	81,000 ~ 614,000	4F, 4E
	35	Dalian Zhoushuizi International Airport	8,587,079			
Small hub airports (24)	36	Shijiazhuang Zhengding International Airport	8,203,974	1.7 million ~ 8.5 million	13,000 ~ 209,000	4E, 4D, and 4C
	59	Weihai Dashuibo Airport	1,807,384			

continued

Category	Ranking	Representative Airport	Passenger Throughput (person-times)	Passenger Throughput Range (person-times)	Terminal Building Area Range (m²)	Airfield Area Grade
Non-hub airports (181)	60	Yancheng Nanyang International Airport	1,691,883	Below 1.7 million	500 ~ 55,700 (excluding Ordos Ejin Horo International Airport) [1]	4E, 4D, 4C, and 3C
	240	Chongqing Xiannvshan Airport	200			

Note: [1] The terminal building area of Ordos Ejin Horo International Airport is 105,300 square meters. Due to the impact of the COVID-19 pandemic in 2020, the passenger throughput of this airport dropped dramatically, so it was not included in the statistics on non-hub airports for that year.

In 2021, China's aviation industry began to recover from the impact of COVID-19 pandemic, with passenger volumes increasing by 5.9% nationally. The total passenger throughput of all civil airports in China reached 907,482,935 person-times. Based on the proposed classification standards, the airports are categorized as follows.

13 large hub airports: Located in Beijing, Shanghai, Guangzhou, Chengdu, Shenzhen, Kunming, Xi'an, Chongqing, Urumqi, Harbin, etc., with annual passenger throughput ranging from 9.07 million to 40.25 million person-times (excluding the newly opened Chengdu Tianfu International Airport), terminal building area ranging from 244,000 to 1.41 million square meters, and airfield area grades of 4F and 4E.

21 regional hub airports: Located in Hangzhou, Shenyang, Jinan, Lanzhou, Nanning, Guiyang, Lhasa, Wenzhou, etc., with annual passenger throughput ranging from 9.07 million to 28.2 million person-times, terminal building area ranging from 81,000 to 614,000 square meters, and airfield area grades of 4F and 4E.

31 small hub airports: Located in Fuzhou, Hohhot, Xishuangbanna, Hulunbuir, etc., with annual passenger throughput ranging from 1.82 million to 9.07 million person-times, terminal building area ranging from 14,000 to 216,000 square meters, and airfield area grades of 4E, 4D, and 4C.

183 non-hub airports: Located in Ganzhou, Dehong, Enshi, etc., with annual passenger throughput below 1.82 million person-times, terminal building area ranging from 500 to 44,000 square meters, and airfield area grades of 4E, 4D, 4C, and 3C. (Table 1-6)

Table 1-6 Civil Airports Classification of China in 2021

Category	Ranking	Representative Airport	Passenger Throughput (person-times)	Passenger Throughput Range (person-times)	Terminal Building Area Range (m^2)	Airfield Area Grade
Large hub airports (13)	1	Guangzhou Baiyun International Airport	40,249,679	9.07 million ~ 40.25 million (taking international passengers into consideration)	244,000 ~ 1.41 million	4F, 4E
	2	Chengdu Shuangliu International Airport	40,117,496			
	3	Shenzhen Bao'an International Airport	36,358,185			
	4	Chongqing Jiangbei International Airport	35,766,284			
	5	Shanghai Hongqiao International Airport	33,207,337			
	6	Beijing Capital International Airport	32,639,013			
	7	Kunming Changshui International Airport	32,221,295			
	8	Shanghai Pudong International Airport	32,206,814			
	9	Xi'an Xianyang International Airport	30,173,312			
	11	Beijing Daxing International Airport	25,051,012			
	18	Urumqi Diwopu International Airport	16,880,507			
	25	Harbin Taiping International Airport	13,502,030			
	47	Chengdu Tianfu International Airport	4,354,758[1]	/		

INTRODUCTION

continued

Category	Ranking	Representative Airport	Passenger Throughput (person-times)	Passenger Throughput Range (person-times)	Terminal Building Area Range (m²)	Airfield Area Grade
Regional hub airports (21)	10	Hangzhou Xiaoshan International Airport	28,163,820	9.07 million ~ 28.02 million	81,000 ~ 614,000	4F, 4E
	33	Wenzhou Longwan International Airport	9,231,409			
Small hub airports (31)	34	Fuzhou Changle International Airport	9,037,195	1.82 million ~ 9.07 million	14,000 ~ 216,000	4E, 4D, and 4C
	65	Hulunbuir Hailar Airport	1,825,229			
Non-hub airports (183)	66	Ganzhou Huangjin Airport	1,808,479	Below 1.82 million	500 ~ 44,000 ②	4E, 4D, 4C, and 3C
	248	Changhai Dachangshandao Airport	164			

Notes: ① Chengdu Tianfu International Airport was newly opened in 2021, so it had a lower passenger throughput that year.
② Due to the same reason with Table 1-5, Ordos Ejin Horo International Airport is not included in this table.

　　Based on the data analysis of these three years, it is evident that while the categories of a few airports may have been adjusted due to changes in annual passenger volumes, the majority of airports saw no change in category. The four-tier classification of airports in the "National Comprehensive Airport System Classification Framework" aligns with the four-level hierarchy of international hubs, regional hubs, artery airports, and regional airports outlined in the "Plan of National Civil Transportation Airport Layout." Due to the impact of the COVID-19 pandemic, the data of 2020 and 2021 does not accurately reflect typical airport throughput of those airports. Therefore, this book defines Chinese airport classification based on 2019 statistics and the classification indicators used above (Table 1-7).

　　Large hub airports (international hub airports): These airports are primarily located in cities of national political, economic, and cultural significance, often serving as major gateway hubs due to their strategic geographical positions as provincial capitals or directly-administered municipalities. At present, China has 13 large hub airports. Before the COVID-19 pandemic, these airports typically handled annual passenger throughput ranging from 40 million to 100 million person-times. Notably, Urumqi Diwopu International Airport and Harbin Taiping International Airport, while handling passengers fewer than 40 million person-times annually, are also classified as large hub airports due to their strategic international locations.

Regional hub airports: These airports are mainly found in provincial capitals and some municipalities with independent planning status, totaling around 23. They primarily function as provincial distribution centers and cross-provincial transfer hubs for domestic passengers. In normal circumstances, their annual passenger throughput ranges from 13 million to 40 million person-times, forming the secondary backbone of China's aviation transportation network.

Small hub airports (artery airports, or mainline airports): Located in prominent tourist cities and prefecture-level cities, these airports handle annual passenger throughput ranging from 2.7 million to 13 million person-times. China currently has approximately 30 airports in this category.

Non-hub airports (regional airports): These airports serve regional centers or specific counties with annual passenger throughput of less than 2.7 million person-times. There are approximately 180 airports of this category in China, accounting for about 75% of the total number of airports nationwide.

Table 1-7 Civil Airport Classification of China Adopted by This Book

Category	Passenger Throughput Range (person-times)	Terminal Scale Range (m^2)	Dominant Airfield Area Grade
Large Hub Airports	40 million ~ 100 million (taking international passengers into consideration)	244,000 ~ 1.41 million	4F
Regional Hub Airports	13 million ~ 40 million	81,000 ~ 614,000	4E
Small Hub Airports	2.7 million ~ 13 million	14,000 ~ 216,000	4E
Non-Hub Airports	Less than 2.7 million	500 ~ 55,700	4C

Note: Data in this table are referenced from the National Civil Transportation Airport Production Statistics Bulletin 2019 released by the Civil Aviation Administration of China and various airport official websites.

Overall, the "regional airports" discussed in this book refer to airports with annual passenger throughput of less than 2.7 million person-times (constituting about 0.2% of China's total aviation passenger throughput), with terminal building area not exceeding 100,000 square meters. These airports typically have an airfield area grade of 4C, primarily accommodating mainline aircraft like the B737, A320, and smaller types.

1.1.3 Scale of Regional Airports and Types of Terminal Buildings

By 2022, regional airports in China have seen a range of terminal building area from 500 to 55,700 square meters, varying in scale but still requiring sound configuration of essential functions. Therefore, studying the functional flows of regional airports, subdividing the scale of terminal buildings based on throughput and service levels, and selecting the most suitable building configuration in consideration of traffic, investment, and airside organization factors are particularly important in the early stages of regional airport construction.

At present, the configuration type of terminal buildings for regional airports mainly include single-story, one-and-a-half-story, and two-story. This book takes 179 regional airports in China in 2019 as examples, summarizing the scale and configuration type of terminal buildings, and proposing subdivision indicators for different regional airport configurations (Table 1-8). The author further divides the terminal building scale and annual passenger throughput of these regional airports into several ranges, as shown in Figures 1-1 and 1-2.

Table 1-8 Scale and Type Statistics of Terminal Buildings of Regional Airports in China (Based on Data of 2019)

Terminal Type	Annual Passenger Throughput (ten thousand person-times)	Terminal Building Scale (ten thousand square meters)	Quantity	Percentage in All Regional Airports
Two-story	16.48 ~ 253.15	1.43 ~ 5.57	17	9.50%
One-and-a-half-story	3.95 ~ 267.91	0.50 ~ 3.2	90	50.28%
Single-story	3.26 ~ 171.12	0.05 ~ 1.4	72	40.22%

Note: The passenger throughput statistics in this table exclude Jiuzhaigou Huanglong Airport, which just resumed flights in 2019, and Ganzi Gesar Airport, which was newly opened at the end of 2019.

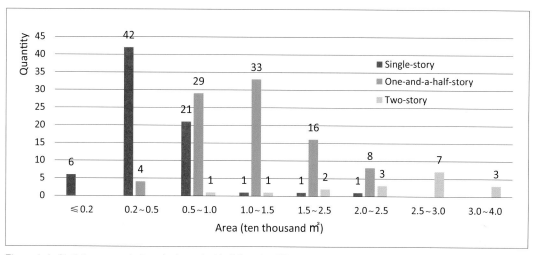

Figure 1-1 Statistics on scale (area) of terminal buildings in different types of China's 179 regional airports in 2019

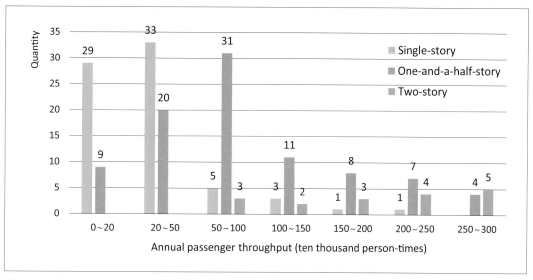

Figure 1-2 Statistics on annual passenger throughput of terminal buildings in different types of China's 179 regional airports in 2019

Data analyses above show that the area of small and medium-sized regional airports in China mainly ranges from 2,000 to 30,000 square meters, with annual passenger throughput of no more than 2.7 million person-times. Specifically, single-story terminal buildings generally have an area of less than 5,000 square meters, with annual passenger throughput of no more than 1.5 million person-times; most one-and-a-half-story terminal buildings have an area of 10,000 to 25,000 square meters, with annual passenger throughput of up to 2.7 million person-times; two-story terminal buildings generally have an area of 15,000 to 40,000 square meters, with annual passenger throughput of 1.5 million to 3 million person-times. It can be seen that the one-and-a-half-story terminal buildings are the most adaptive, due to their high flexibility on area ranging from 2,000 to 25,000 square meters as well as the passenger throughput capacity of up to 2.7 million person-times. Therefore, the one-and-a-half-story configuration takes dominance among the existing terminal buildings of regional airports in China.

Besides, compared to single-story terminal buildings, one-and-a-half-story terminal buildings offer a better indoor boarding experience through near-stand boarding corridors. Additionally, because of the separated arrangement of zones for boarding waiting, check-in, security check, and baggage claim, the one-and-a-half-story terminal building can accommodate more passengers. As for two-story terminal buildings, they inherently have higher demand for area and investment due to the two-story configuration, and require front-building elevated roads for vertical traffic diversion. However, as passenger throughput of new airports often struggle to meet the requirement by such type, it is usually difficult for airports with two-story terminals to get approved. In conclusion, the one-and-a-half-story terminal building, with its excellent compatibility, good user experience, and significant economic advantages, has a positive outlook in the future regional airport development.

1.2 Fundamental Study of One-and-a-Half-Story Terminal Building of Regional Airport

1.2.1 General Characteristics

The one-and-a-half-story terminal building is a concept based on the vertical process flow of airport terminals, characterized by a section consisting of one full story plus a half story. Taking the terminal building section depicted in Figure 1-3 as an example, this type typically includes a full-height hall on the landside, where the ground floor accommodates check-in counters, security checkpoints, and the arrival hall. Facilities adjacent to the airside include baggage claim hall, boarding lounges for remote stands, and various equipment rooms. The second-floor boarding lounge for nearby stands is generally spatially

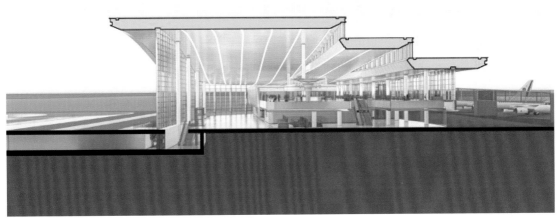

Figure 1-3 Sectional diagram of a one-and-a-half-story terminal building

connected to the landside hall. Under the same roof, they together create an open and spacious indoor environment.

1.2.2 Passenger Flow

Taking the terminal building designed by the author's team for Dingri Airport in the Xizang Autonomous Region as an example (Fig. 1-4), the process flow of a one-and-a-half-story terminal for passengers is straightforward: passengers arrive at the terminal via a ground-level road system, with both departure and arrival halls located on the ground floor. After completing ticketing and security check procedures, departing passengers can ascend to the second-floor boarding lounge via escalators to board through nearby boarding bridges, or wait in the first-floor remote boarding lounge and reach their aircraft via shuttle buses or on foot. Arriving passengers disembark via boarding bridges, descend from the second floor through the arrival corridor to the ground-floor baggage claim hall, then exit the terminal through the arrival hall.

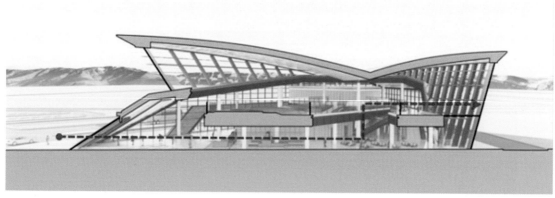

Figure 1-4 Passenger flow in a one-and-a-half-story terminal building

1.2.3 Traffic Organization

The landside traffic organization of one-and-a-half-story terminal building primarily includes external and internal road systems. The external road system serves passengers and typically consists of entrance and exit roads, roads in front of the terminal, curbside roads, and loop redundancy routes. The internal road system is used by the airport's comprehensive support facilities.

Due to the single-level elevation of the terminal building's front road, effectively organizing different traffic streams requires multiple curbside roads. Depending on whether passengers are arriving or departing, curbside roads can be categorized into arrival and departure curbsides. They can also be classified by vehicle type, such as taxi, bus, social vehicle curbside roads, and VIP passenger curbside roads. Following the principle of prioritizing public transport, bus curbsides are generally positioned on the inner upstream lanes, while arrival curbsides for taxis and social vehicles are on the inner downstream lanes.

Curbside typically consists of a vehicular lane, a pedestrian pathway alongside it, and a parking area adjacent to the pathway. This area serves as a waiting zone for passengers boarding or alighting from vehicles, making the length of curbside a crucial design parameter. To prevent traffic disruptions, vehicles generally move in a fixed direction in front of the terminal building, forming a loop traffic flow.

Based on the three aspects above, subsequent chapters in this book will provide a detailed discussion on the design methods of one-and-a-half-story terminal buildings.

1.3 Significance of Design Research on Regional Airport Terminal Building

1.3.1 Practical Significance

Amidst the continuous increase of large hub airports and the ongoing enhancement of related facilities and services, China's civil aviation industry is setting new goals. "The '14th Five-Year Plan' for Civil Aviation Development" outlines a series of objectives for the development of regional airports, underscoring their importance in civil aviation sector. From a national development perspective, the construction of regional airports is pivotal in establishing a modern national airport system. "The National Comprehensive Three-dimensional Transportation Network Plan Outline," released in February 2021, aims to establish approximately 400 civil transportation airports by 2035. This initiative envisions a comprehensive airport system with world-class airport clusters and international aviation (cargo) hubs as its core, supported by regional hub airports and complemented by regional and general airports. Building a modern, convenient, and extensive network of regional airports is imperative for China to strengthen its position as an aviation powerhouse.

Moreover, the development of regional airports aligns with China's policies for the Western Development Strategy and the comprehensive construction of a moderately prosperous society. During "The 13th Five-Year Plan" period, China made historic strides in developing its comprehensive transportation system. Over these five years, the infrastructure network expanded significantly, exceeding 6 million kilometers in total length. "The Ten Verticals and Ten Horizontals" comprehensive transport corridors are completing, while high-speed rail mileage doubled, covering over 95% of cities with populations exceeding one million. The highway system now reaches more than 98% of cities with populations exceeding 200,000. Despite these achievements, disparities in comprehensive transport development persist, particularly in remote western regions, ethnic minority areas, and regions with challenging terrains where contrusting highways and high-speed railways requires high investment and technologies. Therefore, constructing regional airports offers a cost-effective and environmentally friendly solution, effectively stimulating economic development in these regions, leveraging their rich historical, cultural, and natural resources, improving regional connectivity, and enhancing the quality of life for local residents.

In conclusion, faced with significant construction demands, this book aims to provide a framework for the design of regional airports, with a particular focus on one-and-a-half-story terminal buildings. It seeks to systematically summarize and explore innovative design approaches and establish an evaluation system for design innovativeness. The author hopes this book will foster greater consensus among regional airport designers and serve as a valuable reference for future design and construction endeavors.

1.3.2　Theoretical Significance

With the rapid development of China's economy and society, people's living standards continue to rise, along with increased demands for air travel and a growing appreciation for aesthetic experiences. This trend places new requirements on the design of transportation infrastructure. Airports, as crucial public infrastructure, are gaining more attention. Many local governments now fully recognize the role of civil airports in enhancing local prestige, attracting investment, boosting tourism, and driving regional development. In this context, terminal buildings also serve cultural promotion and resource integration.

Currently, research on regional airports in China predominantly focuses on fundamental aspects such as functional and spatial organization. Less attention has been given to the higher demands placed on terminal buildings in the new era and the corresponding need for innovative design methods. This book aims to analyze the latest international and domestic examples of regional airport design, extracting insights from successful cases and dissecting shortcomings. Drawing on the author's extensive experience in airport terminal design, this book aims to establish standards and methodologies for designing excellent terminal buildings, and contribute to the field of innovative airport design research in China.

PLANNING

GENERAL COMPOSITION AND BASIC ELEMENTS OF REGIONAL AIRPORT

2.1 Spatial Composition of Airport

An airport refers to a designated area on land or water used for aircraft takeoff, landing, and ground activities, including associated buildings, installations, and facilities. Similar to general airports, the operational system of a regional airport consists of the airside support system (terminal space), ground support system (airfield area), passenger and cargo service systems (terminal area), and work support systems (work area). (Fig. 2-1)

Airport space (airside support system) includes the space and conditions for flight stopover, such as air traffic control, navigation, communication, and meteorology systems.

Airfield area (ground support system) mainly consists of the movement area and apron. The movement area is the ground space designated for aircraft takeoff, landing, and taxiing, while the apron is the area designated for aircraft taxiing, parking, standing, passenger embarkation and disembarkation, cargo loading and unloading, maintenance, and refueling. Spatially, the airfield area is enclosed by buildings at the airside/landside interface and outdoor isolation facilities.

Terminal area (passenger and cargo service system) refers to the area where the terminal building and its supporting facilities, such as the station apron, transportation systems, and service buildings, are located. This area primarily serves passengers and is centered on the terminal building. The terminal area is

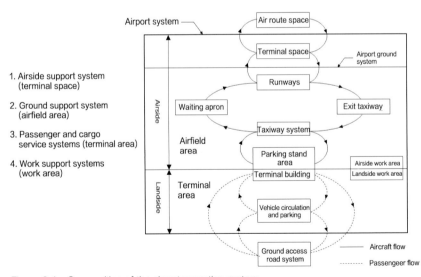

Figure 2-1 Composition of the airport operation system

divided into airside and landside areas. The airside area mainly comprises the airside part of the terminal building and the terminal apron, while the landside area mainly comprises the landside part of the terminal building, the landside passenger departure and arrival flow systems, and the comprehensive service area.

Work area (work support system) mainly refers to the area providing service support for the airfield area and terminal area, including major and auxiliary production facilities, ground transportation, and public facilities.

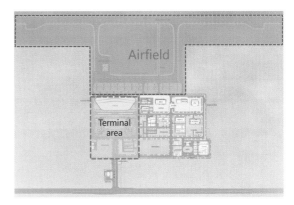

Figure 2-2 Airfield and terminal area of a certain airport

The functions and facilities of the four major areas should be developed synchronously to coordinate with each other, ensuring the safety and efficiency of airport operations. The airfield area and terminal area are regarded as the primary functional areas of a regional airport. (Fig. 2-2)

2.2 Basic Introduction to Airfield Area

This section focuses on the two main components of the airfield area: the movement area (where aircraft operate) and the apron (where aircraft park).

2.2.1 Movement Area

The movement area is the space designated for aircraft takeoff, landing, and taxiing. This book will focus on the runway and taxiway systems.

2.2.1.1 Runway Strip

The runway strip is a designated area that includes the runway, stopway (if provided), and adjacent areas. It is intended to reduce the risk of damage to aircraft if they deviate from the runway, and to ensure the safety of aircraft during takeoff or landing operations. The runway strip typically includes the runway, turn pad, shoulders, and the surface of runway strip.

Runway: A rectangular area designated on land airports, prepared and marked for aircraft landing and takeoff.

Turn pad: A designated area adjacent to the runway, intended for aircraft to complete a 180° turn on the runway.

Shoulder: A prepared area connected to the runway, taxiway, or apron pavement, used as a transition between the pavement and adjacent areas. (Fig. 2-3)

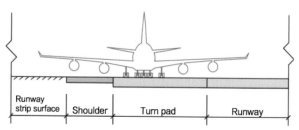

Figure 2-3　Cross-sectional diagram of the runway strip

2.2.1.2　Taxiway

A taxiway is a designated pathway for aircraft to taxi and connect different parts of the airport. Depending on the orientation with respect to the runway, taxiways can be classified as parallel taxiways or connecting taxiways. Depending on functions, taxiways can be categorized into stand taxiways, apron taxiways, rapid exit taxiways, and end-around taxiways.

Stand taxiways: Part of the apron, used solely for aircraft movement into and out of parking stands.

Apron taxiways: Part of the taxiway system but located on the apron, used for aircraft to taxi across or through the apron.

Rapid exit taxiways: Connected to the runway at acute angles, allowing the landing aircraft to quickly enter or exit the runway.

End-around taxiways: Located beyond the end of the runway, used by aircraft to bypass and avoid crossing the runway.

2.2.2　Apron

The apron is a designated area within the airport used for various activities such as passenger embarkation and disembarkation, cargo and mail loading and unloading, refueling, parking, and maintenance of aircraft. The terminal apron, specifically, is a part of the apron located near the terminal building. This section mainly focuses on the basic spatial composition of the parking stands within the terminal apron.

2.2.2.1 Basic Dimensions of Parking Stand

In designing parking stands, two important factors are the basic dimensions of aircraft and the methods of aircraft arrival and departure.

1. Aircraft Basic Dimensions

An aircraft primarily consists of the wings, fuselage, tail assembly, landing gear, and powerplant. (Fig. 2-4)

According to the "Aircraft Classification Table" issued by the International Civil Aviation Organization (ICAO), aircraft are categorized

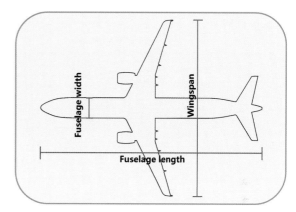

Figure 2-4 Basic dimensions of an aircraft

into Classes A, B, C, D, E, and F based on factors such as speed upon entry to the runway. The sizes of different classes of aircraft vary significantly. For instance, the Cessna AC208, a Class A aircraft, has a fuselage length of 11.4 meters, a wingspan of 15.87 meters, and a height of 4.53 meters. In contrast, the renowned "Superjumbo" Airbus A380, a Class E aircraft, has a fuselage length of 72.75 meters, a wingspan of 79.75 meters, and a height of 24.09 meters.

Due to the lower passenger throughput, regional airports with one-and-a-half-story terminals primarily serve narrow-body aircraft of Class C. Representatives of Class C aircraft include the Boeing 737 series and the Airbus A320. Additionally, China's first independently produced large aircraft, the C919, belongs to Class C with a fuselage length of 38.9 meters, a wingspan of 35.8 meters, and a height of 11.95 meters. (Table 2-1)

Table 2-1 Dimensions of Some Types of Class C Aircraft (in meters)

Aircraft Model	Wingspan	Fuselage Length	Height
Boeing B737-100/200	28.35	29.54	11.28
Boeing B737-300	28.88	33.40	11.13
Boeing B737-600	34.31	< 33.63	12.55

continued

Aircraft Model	Wingspan	Fuselage Length	Height
Boeing B737-700	34.31	33.63	12.55
Boeing B737-800	34.31	39.48	12.55
McDonnell MD-80	32.87	45.06	9.04
McDonnell MD-90	32.87	46.51	9.33
Airbus A320-200	33.91	37.57	11.80
C919	38.9	35.8	11.95
ARJ 21-700	33.47	27.29	8.44

Note: Data is provided by jet bridge manufacturer.

The fuselage length and wingspan of an aircraft impact not only the design of the terminal building's airside, but also the internal functional layout of the terminal. For instance, the length and width of the largest aircraft at an airport determine the configuration of emergency rescue facilities within the terminal.

According to "Emergency Rescue Facilities and Equipment Configuration for Civil Transport Airports" (GB 18040—2019), there are 10 levels for emergency rescue support of airports. These levels are determined by the dimensions of the largest aircraft type the airport serves.

Table 2-2 Classification of Airport Emergency Rescue Support Levels (in meters)

Emergency Rescue Support Level	Maximum Fuselage Length	Maximum Fuselage Width
1	< 9	2
2	9 ~ 12 (excluded)	2
3	12 ~ 18 (excluded)	3
4	18 ~ 24 (excluded)	4
5	24 ~ 28 (excluded)	4

continued

Emergency Rescue Support Level	Maximum Fuselage Length	Maximum Fuselage Width
6	28 ~ 39 (excluded)	5
7	39 ~ 49 (excluded)	5
8	49 ~ 61 (excluded)	7
9	61 ~ 76 (excluded)	7
10	≥ 76	8

Note: Data is sourced from Article 3.2.1 of "Emergency Rescue Facilities and Equipment Configuration for Civil Transport Airports" (GB 18040—2019).

According to Table 2-1, the wingspans of the main aircraft models at small regional airports, such as the B737 and A320, fall within the range of 28 to 39 meters. Therefore, the emergency rescue support level at small regional airports usually does not exceed Level 6.

Once the airport's emergency rescue support level is determined, we can then ascertain the number and size of emergency rescue facilities to be set up within the airport terminal building. (Table 2-3)

Table 2-3 Quantity of Airport Emergency Rescue Institutions

Rescue Institution Category	Emergency Medical Support Level			
	Levels 1 ~ 4	Levels 5 ~ 6	Levels 7 ~ 8	Levels 9 ~ 10
Emergency medical center	—	—	0 ~ 1	1
First aid station	—	0 ~ 1	1	≥ 1, Set according to number of runways
First aid room	0 ~ 1	1	≥ 1, Set according to area of terminal building	

Notes:
① Data is sourced from Article 4.2.5 of "Emergency Rescue Facilities and Equipment Configuration for Civil Transport Airports" (GB 18040—2019).
② For airports of Level 7 or higher, the maximum distance between emergency medical support facilities such as first aid rooms and first aid stations in the passenger concentration area should not exceed 1,000 meters. Otherwise, additional first aid rooms and first aid stations should be set.

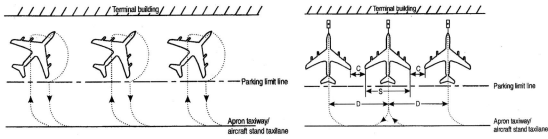

Figure 2-5 Diagram of "taxi-in, taxi-out" mode   Figure 2-6 Diagram of "taxi-in, push-out" mode

Regional airports typically have an emergency rescue level of Level 6. According to Table 2-3, one first aid room should be established, emergency medical center is optional, and there is no requirement for first aid station.

2. Aircraft Arrival and Departure Modes

Aircraft propulsion systems provide forward thrust only. When ground conditions restrict movement, aircraft are towed into and out of bays by tug tractors. Therefore, there are mainly two methods for aircraft operations at parking stands: "taxi-in, taxi-out" and "taxi-in, push-out."

Taxi-in, Taxi-out: In this mode, aircraft use their own power for both arrival and departure. This method is straightforward but requires a larger area. It is typically used for aircraft at remote stands. (Fig. 2-5)

Taxi-in, Push-out: In this mode, the aircraft taxis into the bay using its own power and relies on a tow tug for power when departing the bay. This mode is primarily used for aircraft at near parking stands. (Fig. 2-6)

2.2.2.2 Parking Stand Layout

Based on the earlier discussion, the fuselage length of Class C aircraft typically ranges from 30 to 45 meters, and the wingspan usually falls in between 30 and 35 meters. Considering the dimensions of various Class C aircraft types, the standard size for parking stands at regional airports is generally designed as 45 meters by 36 meters.

Parking stands must also accommodate the spatial requirements for ground services, including cargo handling, refueling, and aircraft maintenance. Ground service vehicles such as tow tugs, catering trucks, cargo loaders, and cleaning vehicles also require sufficient operating space. Therefore, a specified distance must be maintained between parking stands. According to regulations, the minimum safe distance between Class C aircraft stands is 4.5 meters, typically set as 6 meters in practice. Airports with high aircraft movement rates may require a larger safety distance such as 9 meters.

Additionally, the spatial dimensions of near parking stands influence the design of the terminal. With Class C aircraft stands typically being 36 meters wide and a safety distance between stands of 6 meters or 9 meters, the distance between boarding bridges of the terminal building can be estimated at 42 meters or 45 meters. This provides a basis for determining the primary column span of the terminal building: With a 6-meter safety distance, the column span can be calculated as (36 + 6) / 4 = 10.5 meters. For a 9-meter safety distance, the column span would be (36 + 9) / 5 = 9 meters.

Furthermore, the minimum length of terminal building can be estimated based on the number of near parking stands: The number of near parking stands determines the distance between jet bridges at both ends of the building. Adding the length of one column span to this distance provides an approximation of the minimum airside length of the terminal building. (Fig. 2-7)

2.3   Basic Introduction to Terminal Area

This section primarily focuses on the terminal building (including boarding bridges) and the landside transportation system (including parking areas for vehicles) within the terminal area. (Fig. 2-8)

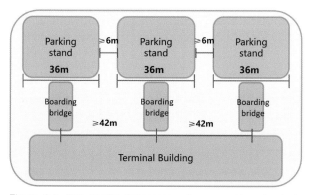

Figure 2-7 The spatial dimensions of near parking stands in relation to terminal building design

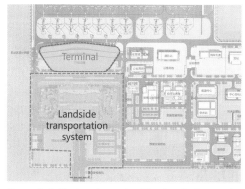

Figure 2-8 Design elements of the terminal area of a certain airport

2.3.1 Terminal Building Configuration

The airport terminal is a building designed to facilitate passenger flow for public transport flights, providing services such as baggage handling, security check, and waiting for boarding. It also serves as the primary interface connecting various ground accesses with passenger-restricted infrastructure systems (such as runway strips and taxiways).

The terminal building can be divided into airside and landside areas, with security checkpoints or international joint inspection zones serving as the boundary between them. The landside area of the terminal connects with the roadway and other landside transportation systems, consisting of ticketing halls, departure and arrival halls, baggage claim halls, VIP lounges, and security checkpoints or international joint inspection halls. Passengers from the landside area go through security check or joint inspection to reach the airside area, and board the aircraft via boarding bridges. The airside area includes commercial zones, waiting areas, boarding bridges, arrival halls, arrival corridors, and baggage handling areas.

The design of the terminal starts with determining its configuration, which should consider factors such as apron layout and landside traffic. Passenger throughput, building scale, and the number of near parking stands are also decisive factors in terminal design.

2.3.1.1 Classification by Connection between Terminal and Airside

Based on the connection between the terminal building and the aircraft stands, terminal configurations can be classified into three main styles: apron-style, finger-style, and satellite hall-style. Finger-style and satellite hall-style configurations are suitable for medium to large and super-large airports with many aircraft stands, but not for small airports. Statistics of China's existing one-and-a-half-story regional airports, with area ranging from 10,000 to 30,000 square meters, show that their configurations are predominantly the apron-style.

Apron-style configuration: Consists of a passenger waiting corridor that can be expanded on both sides, and the main terminal building. The waiting corridor can be linear or take other shapes. Aircraft park on one or both sides of the waiting corridor. (Fig. 2-9)

Finger-style configuration: Consists of a central terminal building and a series of boarding piers. This configuration is commonly used in large and super-large airports with a significant number of aircraft stands, while rarely used in medium-sized airports. Small airports may opt for this configuration due to restricted site conditions or future expansion considerations, which however often involves more complex aircraft operations and lower airside operational efficiency compared to the apron-style. (Fig. 2-10)

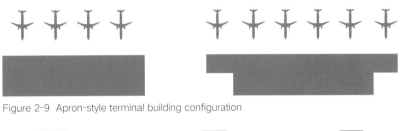

Figure 2-9 Apron-style terminal building configuration

Figure 2-10 Finger-style terminal building configuration

2.3.1.2 Classification by Shape of Terminal Building

For one-and-a-half-story regional airport terminals that adopt apron-style and finger-style configurations, shape of the main building is usually rectangular due to factors such as the number of aircraft stands, aircraft operational efficiency, apron area occupancy, and economic considerations. In addition to the basic rectangular form, other shapes like triangular, circular, and T-shaped forms can also be used.

In a rectangular terminal building (Fig. 2-11), aircraft park on the airside of the terminal building, while the entrance road system and parking lots are organized on the landside.

For triangular terminal buildings (Fig. 2-12), the longer side facing the airside is where aircraft stands are arranged. The two shorter sides facing the landside serve the traffic system, with one side designated for departures and the other for arrivals. This form helps to prevent arrival and departure flows from interfering with each other, and provides a longer curbside compared to the typical rectangular form.

For circular terminal buildings, an example is the famous Charles de Gaulle Airport's Terminal 1 that features an entirely saucer-shaped design, with its first to third floors dedicated to departure flow and the fourth floor serving as the arrival hall, where passengers claim their baggage (Fig. 2-13). Another case is Daocheng Yading Airport, where the terminal building combines rectangular and circular shapes. (Fig. 2-14)

Figure 2-11 The rectangular terminal building of Lancang Jingmai Airport

Figure 2-12 The triangular terminal building of Xizang Dingri Airport

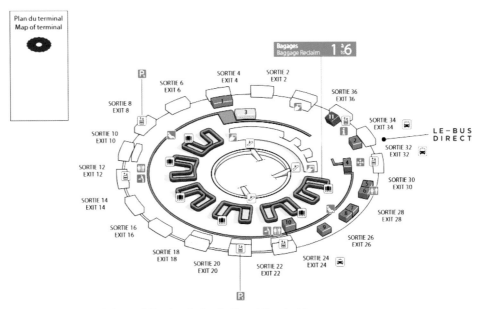

Figure 2-13 The circle form of Charles de Gaulle Airport Terminal 1

Figure 2-14 The circle volume of Daocheng Yading Airport terminal

PLANNING GENERAL COMPOSITION AND BASIC ELEMENTS OF REGIONAL AIRPORT

In T-shaped terminal buildings, a finger-like concourse extends from the airside of the main structure, with aircraft parking on both sides of the concourse. (Fig. 2-15)

2.3.2 Boarding Bridge

The function of boarding bridges is to connect the boarding gates of the terminal building with the aircraft doors. They can be classified into fixed bridges and movable bridges. Movable bridges need to be purchased from boarding bridge manufacturers, while fixed bridges can either be purchased as finished products or designed and constructed as part of the terminal building's functional components.

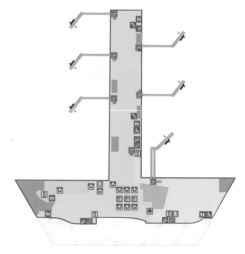

Figure 2-15 The T-shaped terminal building of Beihai Fucheng Airport in Guangxi, China

2.3.2.1 Types of Boarding Bridge

Fixed Bridge: A fixed bridge typically consists of an enclosed gangway, electrical rooms, and evacuation stairs. After passengers pass through the boarding gate inspection, they enter the fixed gangway and then proceed to board through the movable bridge. Fixed bridges can be categorized into single-level bridges and double-decker scissors bridges based on their usage. For one-and-a-half-story terminal buildings, single-level bridges are typically used due to operational considerations.

The minimum clear height of the internal passage in a fixed bridge should not be less than 2,400 millimeters, and the minimum clear width should not be less than 2,200 millimeters. Both lateral sides of the boarding bridge can be made of glass curtain walls or aluminum panel curtain walls, with consideration for smoke exhaust windows. The effective opening area of the outer windows should not be less than 2% of the ground area of the bridge. Sunshade facilities are recommended for the west- and south-facing exterior facades. It is also required to set expansion joints at the junction between the boarding bridge and the terminal building, as well as between the boarding bridge and the apron evacuation stair. (Fig. 2-16)

Movable Bridge: At one end, the movable bridge connects to the fixed bridge via a turntable, while at the other end, it connects to the aircraft door. The movable bridge can adjust its length, height, and angle within a certain range.

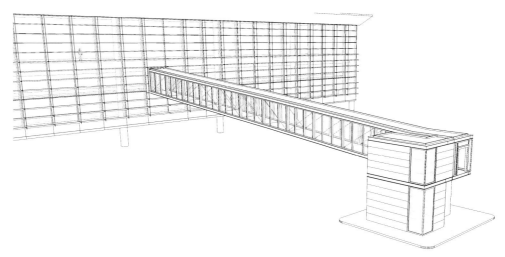

Figure 2-16 Perspective diagram of a single-level fixed bridge in Pudong International Airport satellite hall

Manufactured as a finished product, the movable bridge is subject to constraints such as the aircraft's parking position, the aircraft door, the operating space for service vehicles, and the placement of equipment. As its connection to the fixed bridge is restricted within a certain range of rotation angles and length adjustments, the usage of movable bridges also imposes restrictions and requirements on length and slope of the fixed bridge, as well as position and height of the turntable.

2.3.2.2 Length and Slope of Boarding Bridge

To determine length and slope of the boarding bridge, it is essential to consider the reasonable height for indoor use in the terminal building and the conditions for aircraft docking. Typically, the height of the ground floor of the terminal building is controlled between 5.5 and 6.5 meters.

The height of aircraft doors generally ranges from 2.45 to 3.45 meters (e.g., the lowest door height for a B737-700 is 2.59 meters, and for a C919 is 3.41 meters). Considering that the height of parking positions on the apron is slightly lower than the indoor height of the ground floor of terminal building by about 0.5 meters, when an aircraft docks at a near stand, the relative elevation of the aircraft door to the ground floor of the terminal building is approximately 2.0 to 3.0 meters. Therefore, the height difference between the second-floor departure corridor of terminal building and the aircraft door ranges from about 2.5 to 4.5 meters. This height difference needs to be addressed by the slope of the boarding bridge, making the lowest door height a significant factor in boarding bridge design. Among the common Class C aircraft, the ARJ21-700 has a significantly lower door height (2.26 meters), making it difficult to connect to the terminal building in near parking stands via boarding bridge. For such aircraft, consideration may be given to moving the stop line of stands backward (to increase space for longer boarding bridge), or arranging such aircraft parking only at the remote stands (which should be subject to negotiation with airport operator).

The fixed bridge and the movable bridge are connected via a turntable. The height of turntable is approximately the average of the second-floor height of the terminal building and the height of aircraft door, about 3.75 meters to 4.75 meters, typically set at an elevation of 4 to 4.5 meters.

According to design regulations, the maximum allowable slope for movable bridge and fixed bridge is 1 : 10. From the perspective of usage comfort, the slope should be as gentle as possible if conditions permit, with the recommended slope for fixed bridges not exceeding 1 : 12.

Considering the height difference between the second floor of the terminal building and the turntable, which is approximately 1.5 to 2 meters, the designed slope is usually set at 1 : 12, and the length of the ramp section of the boarding bridge is approximately 18 to 24 meters. At the junction of the fixed bridge and the turntable, there would be a flat buffer section of about 6 meters. Therefore, the minimum length of the fixed bridge should be approximately 24 meters. In practice, the length of fixed bridge is typically set around 30 meters. (Fig. 2-17)

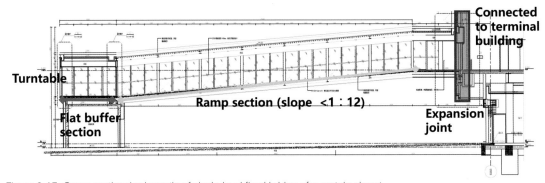

Figure 2-17 Cross-sectional schematic of single-level fixed bridge of a certain airport

In actual projects, it is necessary to communicate closely with design team of the movable bridge to clarify the height of turntable and minimize the length of fixed bridge as much as possible while ensuring reasonable use of the movable bridge.

2.3.2.3 Clear Height under Boarding Bridges

Underneath the boarding bridge there is aircraft head service lane, which allows passage for various service vehicles such as food carts, shuttle buses, and pushback tugs. Additionally, it should also meet the requirements for the passage of fire trucks.

Service vehicles are categorized into small and large vehicles. Small vehicles are those with length not exceeding 6.0 meters, width not exceeding 2.5 meters, and height not exceeding 3.0 meters, while the rest are regarded as large vehicles. The width for a single lane for small vehicles must be no less than 3.5 meters, and the clear height should not be less than 3.5 meters either. For large vehicles, the width for a single lane must be no less than 4.5 meters, and the clear height should be no less than 4.0 meters. According to the "Fire Protection Design Code for Civil Airport Terminal Buildings" (GB 51236—2017), the minimum width and height for passage of fire trucks should both be no less than 4.5 meters.

The general dimensions of some service vehicles are shown in Table 2-4:

Table 2-4 Dimensions of Some Service Vehicles (in millimeters)

Vehicle Type	General Dimensions (Length × Width × Height)
Aircraft towing vehicle	6,180 × 2,220 × 2,130
AC/DC power supply vehicle	5,800 × 2,000 × 2,240
Food cart	4,790 × 2,400 × 3,800
Garbage truck	7,000 × 2,220 × 3,430
Regular passenger stair truck	7,540 × 2,400 × 3,400
Shuttle bus	13,000 × 3,000 × 3,290
Baggage conveyor vehicle	8,800 × 2,200 × 2,050/1,510
Baggage tug	3,860 × 1,670 × 1,720
Baggage cart (tray)	3,100 × 1,500
Pipeline refueling vehicle	8,900 × 2,500 × 3,050

Note: Data is provided by civil aviation operators.

2.3.3 Landside Transportation System

The landside transportation system primarily consists of road system and parking facilities. The road system is further divided into entrance and exit roads, front-of-building roads, curbside lanes, auxiliary service roads, and fault-tolerant loop roads. These components form the main arteries of the terminal area.

The landside transportation system serves as a crucial link for passengers and staff to efficiently enter and exit the terminal area to reach the respective functional buildings. Organizing traffic flows of vehicles and pedestrians, as well as curbside lanes, is key to landside transportation design.

2.3.3.1 Design Outlines of Landside Road System

The focus of landside road system design lies in the organization of entrance and exit roads, front-of-building roads, and traffic flow. Fundamental requirements such as speed control, road width and clear height, road slope, and road turning radius should also be considered.

Speed Control: The speed control requirements for different types of landside roads vary: ① Entrance and exit roads: 60 to 80 km/h; ② Front-of-building roads, curbside lanes, and fault-tolerant loop roads: 20 to 40 km/h; ③ Auxiliary roads: 20 to 50 km/h.

Road Width and Clear Height: The width and clear height of roads vary depending on the types of vehicles using them. For main roads at landside of the terminal area, considering the space requirement of buses, the lane width should be at least 3.5 meters, with a clear height of at least 4.5 meters. If conditions permit, these dimensions can be slightly extended to provide more comfort. According to fire safety regulations for terminal buildings, both the width and clear height of fire truck passages should be no less than 4.5 meters.

Road Slope: According to relevant standards, the horizontal slope of roads is typically between 1.5% and 2%, while the longitudinal slope ranges from 0.3% to 5%.

1. Main Entrance and Exit Roads

The layout of entrance and exit roads determines the overall pattern of landside road system in the terminal area. Depending on the scale

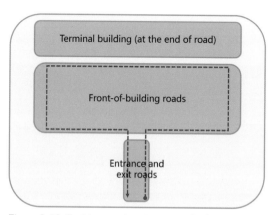

Figure 2-18 End-type main entrance and exit roads

of the terminal area, there are typically three types of main access roads: end-type, through-type, and combination-type.

For small-scale regional airports with lower passenger throughput and smaller terminal areas, the end-type one-way loop roads are the simplest and most efficient, thus are commonly adopted. (Fig. 2-18)

2. Front-of-Building Road

The front-of-building road refers to a one-way loop road in front of the terminal building where passengers gather. As core of the landside traffic pattern, it is crucial for organizing vehicles for passenger drop-off and pick-up, avoiding traffic crossovers, and ensuring efficient connections with parking lots and holding areas.

2.3.3.2 Introduction to Vehicles at Landside

Large buses: These include public transit buses, airport express buses, private-operating buses, tour buses, and long-distance buses. Due to their large passenger volume, large buses require longer time for boarding and alighting, as well as larger parking spaces, thus usually occupying more space of the curbside.

Mid-sized buses: Typically carrying 9 to 20 passengers, including small tour buses, private-operating rental minibuses, and shuttle buses for hotels. Their functions partly overlap with those of large buses, thus they share similar requirements for curbside lanes.

Taxis: Taxis are essential for transporting passengers arriving and departing the terminal area. To ensure efficient service for arriving passengers, terminal areas typically have dedicated lanes and holding areas for taxis.

Other social vehicles: These include private cars and online taxis which constitute a significant proportion of landside vehicles. They usually drop off departing passengers in front of the terminal building and wait in designated parking lots to pick up arriving passengers.

2.3.3.3 Curbside Design

The curbside typically consists of one or more pedestrian paths, temporary parking areas adjacent to the pedestrian paths, and multiple lanes for motor vehicles. It serves as a crucial element in the design of landside road system. The curbside is divided into arrival and departure sections: departure curbsides provide efficient access to the terminal building, while arrival curbsides accommodate passengers waiting for pickups. Effective curbside design should align with the terminal's landside layout to minimize passenger walking distances.

Key considerations for curbside design include: effective length; designated stopping areas; number and width of lanes; widths of pedestrian paths and middle buffer; connection between inner and outer curbsides; and organization of taxi arrival and departure.

1. Classification of Curbsides

The layout of curbsides varies based on the unique travel characteristics of passengers at each airport: By passenger flow process, the curbsides can be divided into departure curbsides and arrival curbsides; by vehicle type, they can be divided into taxi, bus, social vehicle, and VIP vehicle curbsides; by location, they can be divided into inner and outer curbsides (also referred as first curbsides and second curbsides).

2. Parking Modes along Curbside

Parking modes along the curbside include parallel, diagonal, bay-style, and staggered parking. Departure curbsides predominantly utilize parallel parking, allowing vehicles to stop briefly before continuing. Arrival curbsides primarily employ parallel and diagonal parking. Taxi and bus curbsides often employ bay-style and staggered parking, which streamline passenger pickup processes without requiring reversing, thereby enhancing operational efficiency, albeit at the cost of occupying more space.

3. Curbside Settings

To maximize efficiency, curbside lanes are often structured into inner and outer lanes. Following the principle of prioritizing public transport and accommodating high passenger volumes, resources are allocated based on local traffic conditions. Typically, airport express buses utilize inner lanes for both departure and arrival to facilitate efficient boarding and alighting of passengers. Other vehicles use outer lanes for departure. Taxis serving pickups are also designated to outer lanes. Private-operating buses, social vehicles, and online taxis for pickups are usually coordinated in designated parking areas or fixed zones in front of the terminal building. VIP vehicles have dedicated parking areas outside the VIP lounge.

4. Calculation of Curbside Length

The calculation of curbside length for various vehicle types considers their parking duration, passenger load, and the space they occupy along the curbside. This is determined using the following formula:

$$V_i = \sum_{i=1}^{n} P_i \times (\frac{f_i}{N_i} \times \frac{T_i}{3600}) \times C_i$$

Where:

V_i: Required curbside length during peak hours (meters);

P_i: Passenger flow during peak hours;

f_i: Proportion of each vehicle type entering the area;

N_i: Passenger load coefficient of each vehicle type;

T_i: Curbside occupation time of each vehicle type (seconds);

C_i: Length of curbside occupied by each vehicle type (meters).

Empirical data based on extensive research on various transportation modes and passenger behaviors includes:

1) Curbside occupation time (T_i):

Departure curbside: large buses 180 seconds, mid-sized buses 150 seconds, taxis 60 seconds, small social vehicles 90 seconds.

Arrival curbside: large buses 360 seconds, taxis 90 ~ 120 seconds.

2) Passenger load coefficient (N_i):

Large buses: 28 passengers per vehicle; Mid-sized buses: 12 passengers per vehicle; Taxis: 1.5 passengers per vehicle; Small social vehicles: 2.5 passengers per vehicle.

3) Length of curbside occupation (C_i):

Large buses 20 meters, mid-sized buses 14 meters, taxis 8 meters, small social vehicles 8 meters.

Considerations during calculation:

① Adjust P_i based on specific passenger concentration during peak hours at each airport. For instance, at Kunming Changshui International Airport, 40% of passengers arrive within 20 minutes during peak times, while at Hangzhou Xiaoshan International Airport, 65% of passengers arrive within 30 minutes.

② For F_i, approximately 10% of private vehicles drop off passengers in the parking lot, while 90% drop off passengers at the terminal building's departure level.

③ For arrival curbsides, all private vehicles enter the parking lot, so calculations primarily focus on taxis and airport express buses.

2.3.3.4 Vehicle Flow Analysis

At small regional airports with one-and-a-half-story terminal buildings, the landside traffic is often arranged in an end-type, one-way loop road system. Both departure and arrival areas are located on the ground floor at an elevation of 0 meters in the terminal building, resulting in a straightforward and clear passenger flow.

1. Passenger Departure Flow

Passenger departure flow should be efficient and swift. The departuring passengers disembark at the departure curbside lane of the main entrance road, and enter the terminal building promptly. In cases there are two or more curbside lanes, priority is given to public transport vehicles to stop in the inner lanes. (Fig. 2-19)

2. Passenger Arrival Flow

Arriving passengers leaving by taxi need to travel from the terminal building to the taxi curbside lane, while taxis coming from the holding area to pick them up and leave. Arriving passengers leaving by social vehicles need to travel from the terminal building to the designated curbside lane for pickup. Passengers leaving by bus need to proceed to the bus stopping area to board bus and leave. (Fig. 2-20)

3. VIP Vehicle Flow

The VIP lounge is relatively independent, with a separate parking area set outside the entrance / exit. VIP vehicles can enter the VIP parking area from the main entrance and exit roads for passenger pick-up and drop-off. (Fig. 2-21)

2.3.4 Parking Lot and Holding Area

The parking lot serves social vehicles for short-term or long-term parking, while the holding area is primarily for taxis, facilitating centralized management to improve operational efficiency.

2.3.4.1 Parking Lot Design

Parking lots are typically divided into areas for small vehicles and buses. They are usually situated on the ground for cost-effectiveness, though underground facilities can be considered to integrate with

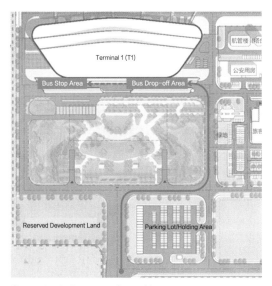

Figure 2-19 Departure flow of bus passengers at a certain airport

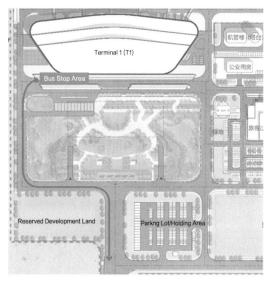

Figure 2-20 Arrival flow of bus passengers at a certain airport

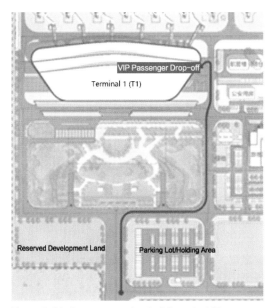

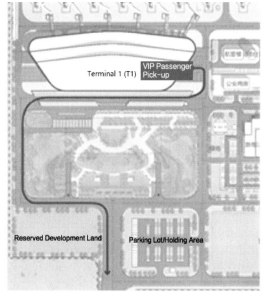

Figure 2-21 Departure and arrival flows for VIP passengers at a certain airport

ground-level landscaping. Underground garages reduce traffic interference but require higher construction costs.

According to the "Civil Aviation Transportation Airport Security Facilities" (MH/T 7003—2017), parking lots or underground parking facilities should not be located within a 50-meter radius around the main terminal building.

When designing social vehicle parking lot, whether to opt for ground or underground facilities depends on site conditions and their relation to the terminal building. Proper placement of entry/exit barriers is crucial for streamlined payment management.

Ground parking lots are typically situated between the main entrance road and the curbside in front of the terminal building. To enhance passenger experience and minimize walking distances to the terminal, the ground parking lot should be as close to the terminal building as possible. However, due to safety requirements under civil aviation regulations, the distance between the ground parking lot and the terminal building must not be less than 50 meters.

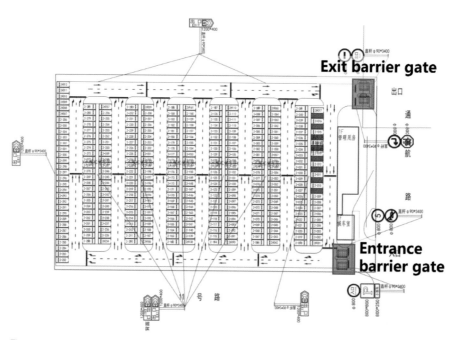

Figure 2-22 Parking lot gate configuration at a certain airport

For underground parking garage, the barrier gates, along with payment systems, are usually positioned at the ramps leading to its entrances and exits (Fig. 2-22). Internal vehicle flow in the garage typically follows a one-way circulation pattern. Moreover, underground parking facilities must also comply with regulations stipulating that its distance from the terminal building should not be less than 50 meters.

In designing parking lots, aside from ensuring efficient internal flow, attention should be given to fire compartmentalization and the placement of evacuation staircases. To enhance the parking lot's quality and reduce investments in mechanical and electrical systems, skylights, patios, and courtyards can be incorporated to improve natural lighting and ventilation.

2.3.4.2 Holding Area Design

The holding area is divided into bus holding areas and taxi holding areas, with design considerations primarily focusing on their relative location to the terminal building, layout, and vehicle flows for entry and exit.

For example, the taxi holding area should be positioned upstream of the traffic flow towards the terminal building, ensuring its internal traffic flow aligns with the overall vehicle flow in front of the terminal. Depending on the method of taxi release, taxi holding areas can be categorized into single-lane release, grouped release(Fig. 2-23), and call-for release systems.

In the holding area, taxis enter the terminal area via the main entrance road, and drop off passengers at the departure curbside lane in front of the terminal building. Afterward, they have the option to exit the terminal area directly or proceed through a one-way-loop road to reach the taxi holding area, where they wait for release. Once released, they swiftly move to the taxi arrival curbside lane to pick up passengers and leave. (Fig. 2-24)

2.3.4.3 Parking Stall Design

The design of parking stalls should be tailored to various vehicle sizes, airport operational requirements, and accessibility standards, categorized as follows:

Standard Parking Stall: Intended for small motor vehicles, with a minimum size requirement of 2.4 meters × 5.1 meters. To enhance passenger comfort, these stalls typically measure around 3 meters × 6 meters and are concentrated in parking lots or garages.

Accessible Parking Stall: Positioned nearest to the curbside, these stalls feature an adjacent aisle at least 1.2 meters wide for wheelchair passage. Access from the wheelchair passage aisle to the curbside

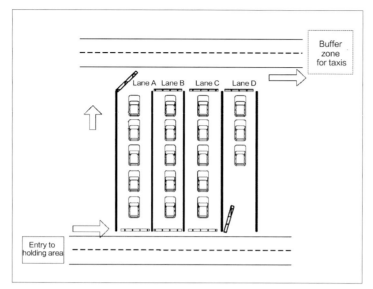

Figure 2-23　Grouped release taxi holding area

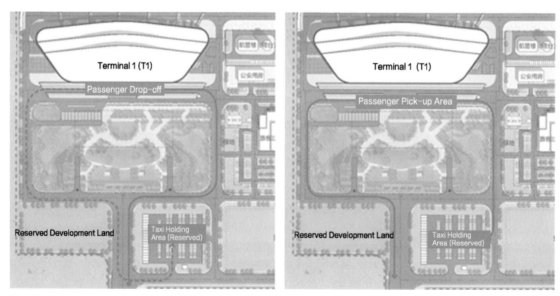

Figure 2-24　Pick-up and drop-off flow of taxis at a certain airport

should be safe, direct, and convenient.

Overnight Parking Stall: Found in parking lots or garages, these stalls are designated for vehicles requiring long-term parking.

Bus Parking Stall: Designed for mid-sized buses (approximately 3.5 meters × 12 meters) and large buses (around 3.5 meters × 16 meters). Short-term parking stalls for airport express buses are typically located in front of the terminal building, while long-term parking for buses is provided in designated bus parking lots.

VIP Parking Stall: Located near the entrance/exit of the VIP lounge, these stalls are exclusively reserved for VIPs.

FUNCTIONS

FUNCTIONAL COMPOSITION AND DESIGN OUTLINES OF TERMINAL BUILDING

3.1 Functional Areas of Terminal and Passenger Flows

A one-and-a-half-story terminal building is primarily formed by various functional areas such as the arrival/departure hall, check-in area, security checkpoint, boarding lounge, baggage claim hall, baggage handling room, and VIP lounge. These functional areas are interconnected synergistically, balancing the relationships between each others. By integrating with the overall operational processes, they collectively form a complete terminal layout that meets the operational requirements of the terminal. (Fig. 3-1)

3.1.1 Arrival/Departure Hall

In the one-and-a-half-story terminal building, the arrival and departure functions are located within the same space. The arrival/departure hall needs to seamlessly connect passenger arrivals and departures with the landside transportation system and complete the check-in process in the departure hall area. The layout and spatial dimensions of the arrival/departure hall directly influence passengers' first impressions of the terminal building. (Fig. 3-2)

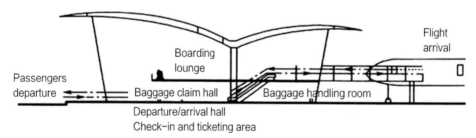

Figure 3-1 Functional composition of a one-and-a-half-story terminal building

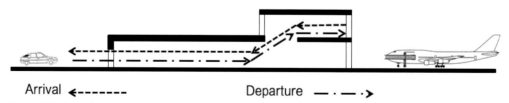

Figure 3-2 Functional flows in the arrival/departure hall

The departure hall primarily serves departing passengers and provides check-in services. The main facilities include various service inquiry counters, seating areas, and landside retail spaces. The arrival hall caters to arriving passengers and those welcoming them. Its main facilities include information boards about arriving flights, arrival gates, connections to urban transportation, and services such as retail and dining, car rentals, hotels, baggage storage, etc. (Fig. 3-3)

The scale of the arrival/departure hall is determined based on the number of users. According to the service standard proposed by the International Air Transport Association (IATA), a Class-C arrival/departure hall should provide an average of 2.0 square meters per person. Thus, the fundamental area of the arrival/departure hall can be calculated. In actual design, the space required for check-in, security checks, passenger passage, and various auxiliary services should also be considered. Finally, the overall configuration of the terminal building is taken into account to determine the width and depth of the arrival/departure hall. The following sections will analyze the basic space requirements of each functional zone. (Fig. 3-4)

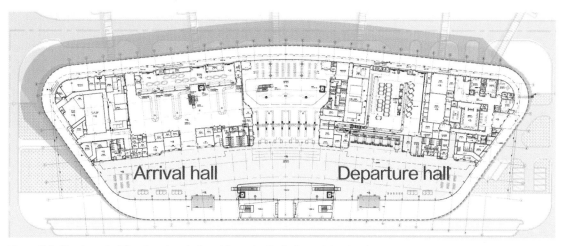

Figure 3-3 Plan layout of the departure hall and the arrival hall of a certain airport

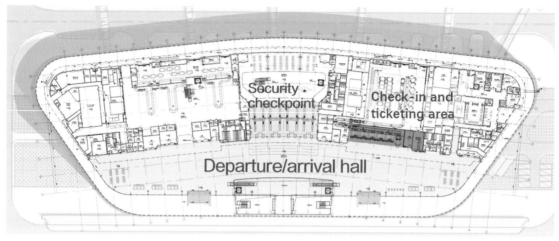

Figure 3-4 Functional zoning of the arrival/departure hall of a certain airport

3.1.2　Check-in and Ticketing Area

　　Adjacent to the departure hall, the check-in and ticketing area allows passengers to conduct flight procedures and check their baggage. It can be divided into several interconnected or independent sections, including domestic, international/regional (this book only focuses on the domestic section), VIP sections, and those for special route flights. The check-in and ticketing area consists of check-in counters, oversize baggage check-in counters, baggage inspection rooms, baggage storage, certificate application, and other functional spaces. (Fig. 3-5)

　　The spatial dimensions of the check-in and ticketing area are determined by the number and layout of counters and self-service check-in equipment, the distance to the ticketing waiting area, and the sizes of various functional rooms.

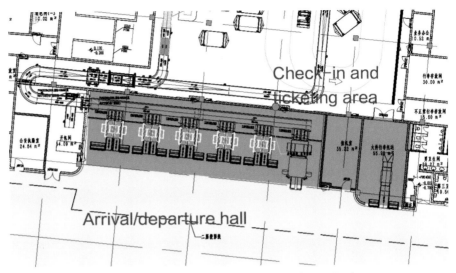

Figure 3-5 Functional layout of the check-in and ticketing area of a certain airport

3.1.3 Security Checkpoint

To ensure aviation safety, departing passengers must undergo security checks for themselves and their carry-on baggage before entering the security control area. The security checkpoint mainly includes the security check passages, security offices, special inspection rooms, and prohibited items storage rooms. (Fig. 3-6)

The number of security check passages, the width and depth of passages, and the type and layout of security equipment together determine the spatial dimensions of the security checkpoint. Typically, the width of a single security check passage ranges from 4.5 to 5.0 meters.

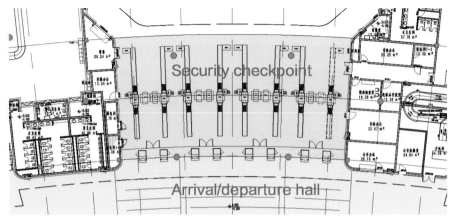

Figure 3-6 Functional layout of the security checkpoint of a certain airport

3.1.4 Boarding Lounge

The boarding lounge is the area where passengers wait for boarding and enjoy relevant services after passing through security checks. It usually includes several boarding gates, seating areas, passenger passages, and facilities for passenger services and retail. (Fig. 3-7)

The spatial dimensions of the boarding lounge are determined based on the layout of the boarding gates. Generally, the boarding gates are arranged along the airside interface. With the average area per person required for a Class-C boarding lounge specified by IATA, and related data on passenger usage, the fundamental area and depth of the boarding lounge can be calculated. In actual design, the space requirements for boarding queues, seating areas, passenger passages, and other functional services should be comprehensively considered to figure out the dimensions of the boarding lounge.

3.1.5 VIP Lounge

The airport provides two types of VIP lounges: the VIP lounges at the landside (referred to hereafter as "landside VIP lounge"), and those for passengers of first-class and business-class located within the boarding lounge (referred to hereafter as "two-class VIP lounge" hereafter).

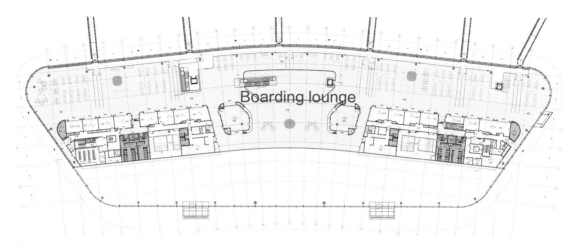

Figure 3-7 Functional layout of the boarding lounge of a certain airport

The basic functional zones of the landside VIP lounge include reception center, VIP hall, VIP rooms, and VIP security checkpoint. The two-class VIP lounges can be managed by airline companies or operated directly by the airport.

The dimensions of each functional space in the VIP lounge should ensure rational use and comfort. The usable area per capita in the VIP lounge should be at least 4 square meters. (Fig. 3-8)

3.1.6 Arrival Corridor

The arrival process can be implemented via a mixed mode (where arrival and departure flows are mixed) or a seperated mode (where arrival and departure flows do not interfere with each other). In the terminal buildings of regional airports, due to limited space, the distance from the arrival gate to the escalators in the baggage claim area cannot support a coherent flow. Adopting the mixed-flow mode might cause congestion in the terminal, disturbing waiting and boarding processes. Therefore, the separated-flow mode is preferred, where passengers are directed through arrival corridors to the baggage claim area. (Fig. 3-9)

The layout of the arrival corridor and boarding gates affects the dimensions of the arrival corridor. Its width should be determined based on passenger behavior, able to accommodate at least two streams

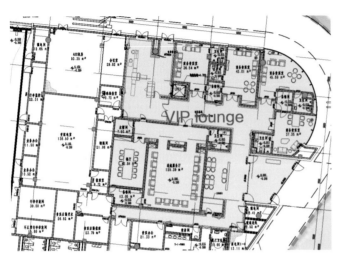

Figure 3-8 Functional layout of the VIP lounge of a certain airport

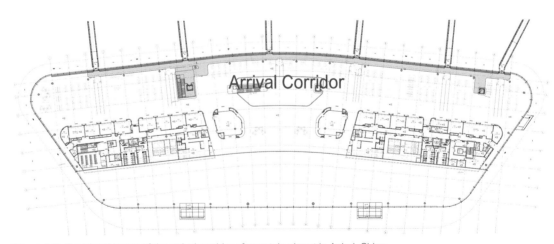

Figure 3-9 Functional layout of the arrival corridor of a certain airport in Anhui, China

of passengers with carry-on baggage while one stream to overtake the other. Typically, it is designed to be 3.0 to 4.0 meters wide.

3.1.7 Baggage Claim Hall

Due to the limited scale of regional airports, the arrival corridor may not effectively divert arriving passengers. Therefore, the baggage claim hall serves not only as the zone for retrieving checked baggage, but also as the first diversion space where passengers can stay for a longer time upon arriving at a terminal.

The basic functional spaces of a baggage claim hall include baggage carousels, oversized baggage pickup areas, baggage inquiry counters, restrooms, etc. (Fig. 3-10)

The spatial dimensions of the baggage claim hall are determined by the number and layout of baggage carousels and the space requirements for passenger use.

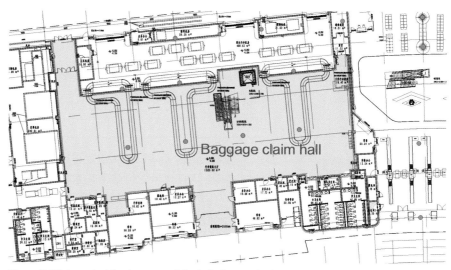

Figure 3-10 Layout of the baggage claim hall of a certain airport in Anhui, China

The number of baggage carousels can be estimated based on peak-hour flight arrivals. Typically, one set of carousels can handle baggage retrieval for 3 Class-C aircraft, or 2 Class-D aircraft, or 1 Class-E aircraft per hour. For one-and-a-half-story terminals in regional airports (annual passenger throughput usually less than 2 million), serving 6 Class-C aircraft can be used as a reference, and 1 to 2 sets of baggage carousels can be considered.

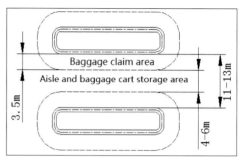

Figure 3-11 Layout of the baggage claim hall of a certain airport

To meet passengers' usage requirements, the spacing between two baggage carousels can be controlled within 11 to 13 meters. This includes a retrieval area around each carousel approximately 3.5 meters wide and a passenger passage of 4 to 6 meters wide between the carousels. (Fig. 3-11)

3.1.8 Baggage Handling Area

The baggage handling area is divided into the outbound baggage handling area and the inbound baggage handling area. The departure baggage handling process is relatively complex, involving tasks such as weighing, security screening, conveying, sorting, and monitoring of passengers' checked baggage. It can be broadly categorized into baggage security screening, baggage collection and transfer, baggage sorting, and baggage loading. The main tasks of arrival baggage handling include baggage sorting, conveying, and addressing irregular baggage.

The layout of the baggage handling area is determined by the overall arrangement of baggage handling systems, facilities for baggage loading and unloading, baggage carts, and other related equipment. The height of the baggage handling room mainly depends on the layout of the baggage system, which should be synchronized with the architectural design of the terminal building. Manual sorting and comprehensive sorting are the two typical methods employed in baggage handling. (Fig. 3-12)

The layout and size of baggage loading and unloading equipment (such as turntables, chutes, and conveyors) can significantly impact the spatial dimensions of the baggage handling area. Additionally, the traffic flow of baggage carts is an important factor in designing baggage handling area.

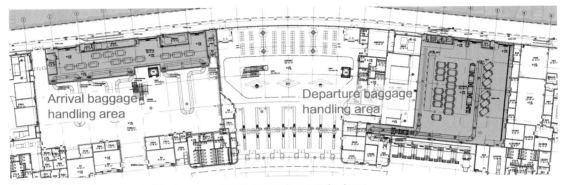

Figure 3-12 Functional layout of the baggage handling areas in a certain airport

3.1.9 Summary on Process Flows in Terminal

3.1.9.1 Domestic Passenger Flows

1. Domestic Departure Flow

Boarding at Near Parking Stands: Departure curbside → Departure hall (buying ticket and insurance) → Ticketing (check-in, security check for checked baggage) → Security check (for passengers and carry-on baggage) → Arrive at boarding gates of near parking stands via escalator → Arrive at boarding lounge on the second floor → Boarding pass check → Boarding via jet bridge.

Boarding at Remote Parking Stands: Departure curbside → Departure hall (buying ticket and insurance) → Ticketing (check-in, security check for checked baggage) → Security check (for passengers and carry-on baggage) → Arrive at boarding lounge for remote stands → Boarding pass check → Boarding at boarding gates for remote stands. (Fig. 3-13)

2. Domestic Arrival Flow

Arrival at Near Parking Stands: Disembarkation → Arrival corridor → Retrieve baggage in baggage claim hall → Baggage tag inspection → Arrival hall → Arrival curbside.

Arrival at Remote Parking Stands: Disembarkation → Retrieve baggage in baggage claim hall → Baggage tag inspection → Arrival hall → Arrival curbside. (Fig. 3-14)

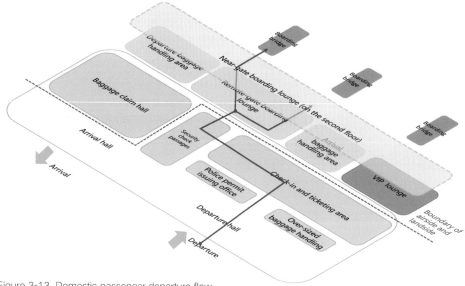

Figure 3-13 Domestic passenger departure flow

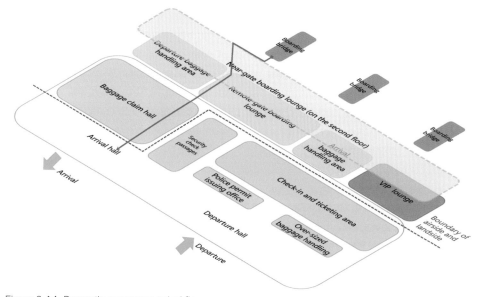

Figure 3-14 Domestic passenger arrival flow

3. Domestic Transfer and Transit Flow

Transfer for Connecting Flight/Transit: Disembarkation → Arrival corridor → Staff inspection → Arrive at boarding lounge on the second floor → Boarding pass check → Boarding via jet bridge. (Fig. 3-15)

Non-Connecting Transfer: Disembarkation → Arrival corridor → Retrieve baggage in baggage claim hall → Baggage tag inspection → Arrival hall → Departure hall (buying ticket and insurance) → Ticketing (check-in, security check for checked baggage) → Security check (for passengers and carry-on baggage) → Arrive at boarding gates of near parking stands via escalator → Arrive at boarding lounge on the second floor → Boarding pass check → Boarding via jet bridge. (Fig. 3-16)

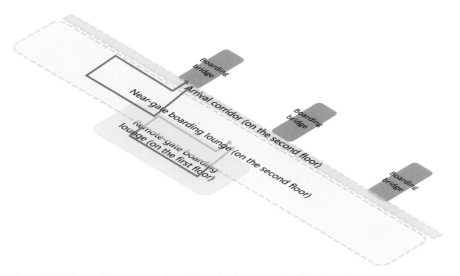

Figure 3-15 Domestic passenger flow of transfer for connecting flight/transit

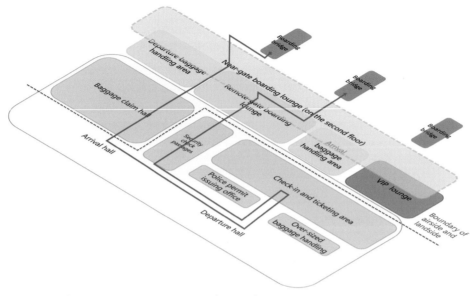

Figure 3-16 Domestic passenger non-connecting transfer process

3.1.9.2 International Passenger Flows

Even though the one-and-a-half-story terminal building of regional airports is small, there is potential of developing international routes or upgrading to an international airport. Therefore, it is necessary to consider international passenger flows in the terminal design for regional airports.

1. International Departure Flow

Boarding at Near Parking Stands: Departure curbside → Ticketing at the departure hall → Customs → Security check → Border control verification → Airside central commercial area → Boarding lounge → Boarding via jet bridge.

Boarding at Remote Parking Stands: Departure curbside → Ticketing at the departure hall → Customs → Security check → Border control verfication → Airside central commercial area → Escalator → Boarding lounge for Remote stands → Boarding via shuttle bus.

2. International Arrival Flow

Arrival at Near Parking Stands: Disembarkation via jet bridge → International arrival corridor → Customs (inspection and quarantine) → (Landing visa hall) → Border control verification → Baggage claim hall → Customs → Baggage tag inspection → Arrival hall → Arrival curbside.

Arrival at Remote Parking Stands: Disembarkation via shuttle bus to arrival gate of remote stands → International arrival corridor → Customs (inspection and quarantine) → (Landing visa hall) → Border control verification → Baggage claim hall → Customs → Baggage tag inspection → Arrival hall → Arrival curbside.

3.1.9.3 Staff Flows

Crew Staff Flow: In the normal operation of regional airport terminals, departure crew staff follow the same process as departure passengers, usually using the staff passage of the main flow or VIP passage. Arrival crew staff can use the dedicated staff passage or enter the arrival area through the departure process after entering the arrival floor.

Airport Staff Flow: The office areas for airport staff are mainly divided into landside offices and airside offices. Airport staff often use the staff passage of the main flow or the cargo passage of the terminal building.

3.1.9.4 Terminal Cargo Handling Flows

1. Dedicated Cargo Flow within Terminal

The flow of dedicated cargo process within the terminal are often coordinated with garbage disposal flow. These process flows usually operate during non-flight hours at day and night, used by security check staff for cargo inspection and garbage transportation. These flows are not used separately during normal operating hours of the airport. (Fig. 3-17)

2. Shared Cargo Flow within Terminal

For terminals with relatively simple functional processes, small area, and low throughput, cargo routes can be combined with passages for staff security checks within the passenger security checkpoint. One dedicated passage should be allocated for staff/returning passengers, and the timing of cargo passing through the terminal security equipment should be managed specifically.

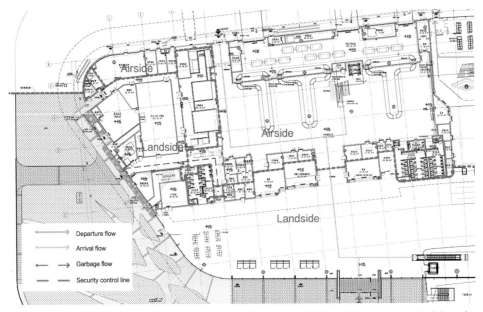

Figure 3-17 Layout of dedicated routes for cargo/garbage transportation within the terminal building of a certain airport in Anhui, China

3.2 Design Outlines of Arrival/Departure Area

3.2.1 Dimensions of Entrance Gate Porch

3.2.1.1 Pre-Security Check Area at Entrance

Before entering the landside area from the landside transportation systems, passengers and their belongings must undergo security checks to ensure the safety of the landside area of the airport. Since the outbreak of the COVID-19 pandemic in 2020, there has been a heightened focus on preventing global infectious diseases. This makes it particularly important to design the curbside buffer area and arrange queueing for pre-security checks based on both the scale of the human body and fundamental

Figure 3-18 The relationship between the curbside, buffer zone, and the landside of the terminal building of Shanghai Hongqiao International Airport Terminal 1

transportation demands.

The area in front of the terminal entrance is a crowded space (Fig. 3-18), and the buffer area between the curbside and landside areas of the terminal must be considered in the flow planning of the entrance area.

Reasonably planning of the buffer area allows for flexible organization of temporary inspections and temperature checks, while also accommodating facilities like canopies along the curbside to organize passenger flow and relieve traffic pressure at the landside. It is suggested that the distance between passengers in pre-security checkpoint queues should be maintained at 1 meter, while ensuring each passenger has a space of 1.2 meters × 1.2 meters for movement. Furthermore, considering the space

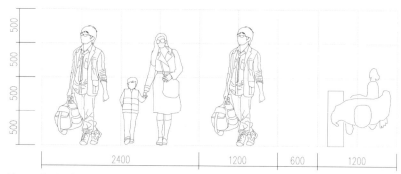

Figure 3-19 Space dimensions for passengers in pre-security checkpoint queues (in millimeters)

demand for a family of three, the width of the queue space should be at least 2.4 meters; for the use of two streams of people simultaneously, the width should be at least 3.0 meters. (Fig. 3-19)

3.2.1.2 Explosive Inspection Area at Entrance

Explosive inspections involve collecting samples from baggage using test paper to detect whether there are explosive particles (nitrate ions), thus to determine if there is an explosion risk. To be more efficient, explosive inspections are often conducted on multiple passengers at once, causing passengers waiting for inspection and those waiting for test results to briefly stay in the inspection area.

Explosive inspections can be conducted in areas designated by soft partitions within departure and arrival halls, or at the entrance gate of the terminal building. The design of this area should also consider the reasonable width for passenger movement.

3.2.2 Planar Dimensions of Departure/Arrival Hall

The planar dimensions of the departure/arrival hall depend on the number of passengers using them, which can be calculated via the following equations:

Number of departure hall users = [passenger count for domestic departure during peak hour × (domestic concentration coefficient + domestic arrival/departure ratio) + passenger count for international departure in peak hour × (international concentration coefficient + international arrival/departure ratio)] × 0.5 + number of approved staff members

Number of arrival hall users = (passenger count for domestic arrivals in peak hour × domestic concentration coefficient + passenger count for international arrivals in peak hour × international concentration coefficient) / 6 + passenger count for domestic arrivals in peak hour × domestic arrival/departure ratio + passenger count for international arrivals in peak hour × international arrival/departure ratio + number of approved staff members

The number of passengers using the departure/arrival hall determines the designed area of the hall, the evacuation width, and the number of evacuation doors. As the departure/arrival hall usually also accommodates people evacuating from other areas, the actual number of evacuation doors in the departure/arrival hall may differ from the calculation results. Table 3-1 lists the usage area and evacuation width of the departure/arrival hall in terminals of different scales.

Table 3-1 Number of Evacuation Doors in Departure/Arrival Hall of Terminals with Different Scales

Annual Passenger Throughput (10,000 person-times)	Usage Area Per Capita (m^2)	Number of Users in Peak Hour	Usage Area (m^2)	Evacuation Width (m)	Number of Evacuation Doors (Taking 1.4-meter-wide door as an example)
50	2.0 (IATA Service Standard C)	262	524	⩾ 2.8	2
100		523	1,046	⩾ 5.6	4
150		785	1,570	⩾ 8.2	6
200		1,046	2,092	⩾ 11	8

Through the study of the plan dimensions of terminals with an annual passenger throughput of around 2 million person-times, it was found that the actual area of the departure hall is close to the calculated capacity value (i.e., the product of the number of users and the usage area per capita). The built area of the departure hall typically constitutes 10% to 17% of the total built area of the terminal (Table 3-2), while the built area of the arrival hall generally makes up 4% to 9% of the total built area of the terminal (Table 3-3).

Table 3-2 Ratio of Departure Hall Built Area to Total Built Area of Terminal of Several Airports

Airport Name	Annual Passenger Throughput (ten thousand person-times)	Area of Terminal Building (m^2)	Width/ W (m)	Depth/ D (m)	Area of Departure Hall (m^2)	W/D	Area Ratio of Departure Hall
Shennongjia Airport, China	2	2,829.55	34.3	12.8	439.04	2:1	15.5%
The new airport of Georgia	No data	4,657.84	40	10	400	4:1	8.6%
Xizang Dingri Airport, China	25	6,612	55	13	715	4:1	10.8%
Shangrao Sanqingshan Airport, China	50.01	9,774.1	96.5	17	1,640.5	5:1	16.8%
Yueyang Sanhe Airport, China	81.1	6,168.1	40	18	720	2:1	11.7%
Cangyuan Washan Airport, China	100	10,712.9	80.9	17.3	1,399.57	4:1	13.1%
Nelson Airport, New Zealand	120	6,304.48	54	17.2	928.8	3:1	14.7%
Lancang Jingmai Airport, China	135	11,774.86	72.2	16.9	120.18	4:1	10.2%
Hulunbuir Hailar Airport, China	255.84	14,746.84	115.5	21	2,425.5	5:1	16.4%

Table 3-3 Ratio of Arrival Hall Built Area to Total Built Area of Terminal of Several Airports

Airport Name	Annual Passenger Throughput (ten thousand person-times)	Area of Terminal Building (m^2)	Width/ W (m)	Depth/ D (m)	Area of Arrival hall (m^2)	W/D	Area Ratio of Arrival Hall
Shennongjia Airport, China	2	2,829.55	12.3	10.6	130.38	1:1	4.6%
The new airport of Georgia	No data	4,657.84	20	12	240	1:1	5.2%
Xizang Dingri Airport, China	25	6,612	39.7	13	516.1	3:1	7.8%
Shangrao Sanqingshan Airport, China	50.01	9,774.1	27.7	17	470.9	1:1	4.8%
Yueyang Sanhe Airport, China	81.1	6,168.1	30	18	540	1:1	8.8%
Cangyuan Washan Airport, China	100	10,712.9	44.9	17.3	776.77	2:1	7.3%
Nelson Airport, New Zealand	120	6,304.48	25.2	17.2	433.44	1:1	6.9%
Lancang Jingmai Airport, China	135	11,774.86	72.2	13	938.6	5:1	8.0%
Hulunbuir Hailar Airport, China	255.84	14,746.84	42	21	882	2:1	6.0%

3.3 Design Outlines of Check-in and Ticketing Area

3.3.1 Spatial Layout of Check-in and Ticketing Area

The check-in and ticketing area is the first functional space that most passengers encounter after entering the terminal building, making its spatial scale particularly important. This area accommodates ticketing services and passenger classification and verification. During times of epidemic prevention, the check-in and ticketing area experiences higher concentration of people during peak hours and longer processing time.

Located near the airside, the departure baggage handling room is usually adjacent to the check-in and ticketing area. The location selection of the check-in and ticketing area is closely related to that of the departure baggage handling room, with terminals of different passenger throughputs adopting different layouts.

3.3.1.1 Centralized Layout of Baggage Handling Rooms

Terminals with lower passenger throughput often opt for a centralized layout, where the baggage handling rooms for departures and arrivals are ideally connected. The departure baggage handling room is positioned behind the arrival baggage handling room along the proceeding direction. This layout ensures a continuous process and minimizes the number of staff required to manage the baggage handling rooms, facilitating the orderly operation of the terminal.

For regional airports with small-sized terminal buildings and limited check-in ticketing equipment, the centralized layout provides a shorter distance from the main entrance to the central axis of the check-in and ticketing area, avoiding significant pedestrian crossings. Self-service check-in passengers can bypass the check-in area and directly enter the security check area. (Fig. 3-20)

3.3.1.2 Distributed Layout of Baggage Handling Rooms

The distributed layout is suitable for regional airport terminals with higher passenger throughput or underground transportation connections. In this layout, the baggage handling rooms for departures and arrivals are separated into two independent areas by the security check area and remote boarding lounge. The departure baggage handling room is positioned behind the arrival baggage handling room along the proceeding direction, ensuring continuity in the process but requiring two sets of staff to manage the baggage handling rooms. (Fig. 3-21)

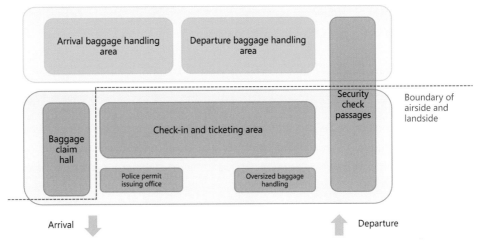

Figure 3-20 Centralized layout of baggage handling rooms

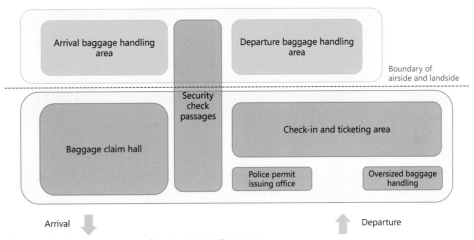

Figure 3-21 Distributed layout of baggage handling rooms

3.3.2 Processes and Constituent Elements of Check-in and Ticketing Area

3.3.2.1 Check-in and Ticketing Process

Traditionally, check-in is conducted at counters on check-in islands. Check-in methods include non-open (each airline has its own check-in counters serving only its passengers) and open (all airlines share all the check-in counters). Currently, most airlines in China adopt the non-open check-in method, often leading to uneven pressure among different check-in counters. Consequently, open self-service check-in equipment has emerged to address this issue.

For passengers without checked baggage, most airports have already provided self-service check-in kiosks to alleviate pressure on the traditional check-in counters. For passengers with checked baggage, the open self-service baggage check-in system is becoming popular, enabling real self-service check-in.

Additionally, an emerging online check-in method should be widely promoted. Passengers can complete check-in procedures online via the internet and print electronic boarding passes in advance. Upon arriving at the terminal, they can directly scan the self-printed boarding passes or ID cards to enter the security checkpoint. This method not only saves time spent in queueing and check-in operations and eliminates the risk of losing boarding passes, but also advocates for environmentally friendly paperless operations, representing a simplified future check-in process. (Fig. 3-22)

3.3.2.2 Quantity of Check-in and Ticketing Facilities

According to the requirements for the functional layout of terminal buildings, the minimum number of check-in counters is six, including two traditional counters (Fig. 3-23), two self-service counters (to meet the requirements for "safe, efficient, green, and harmonious" civil airports in China) (Fig. 3-24), one counter for upper-class passengers, and one duty officer counter (which can also serve as an accessible counter). For a one-and-a-half-story terminal building of a regional airport with less annual passenger throughput ($\leqslant$ 2 million person-times) and high flight concentration rates, the number of counters to be deployed can be calculated as follows:

Number of traditional check-in counters (including self-service baggage check counters) = passenger volume in peak hour × concentration rate × 50% (departure/arrival ratio) × check-in speed / 3600

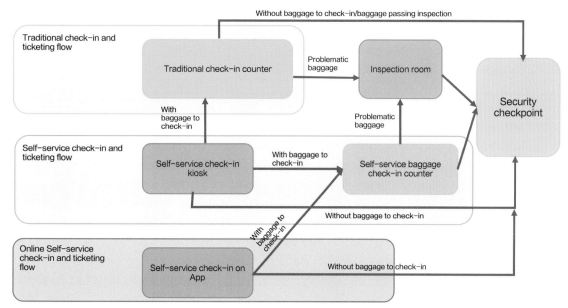

Figure 3-22 Check-in and ticketing process

Figure 3-23 Counters for traditional check-in and ticketing in Terminal 2, Shanghai Hongqiao International Airport

Figure 3-24 Facilities of self-service check-in and ticketing in Terminal 2, Shanghai Hongqiao International Airport

Number of self-service check-in counters = peak hour passenger volume × concentration rate × 50% (departure/arrival ratio) × check-in speed/3600

Traditional check-in and ticketing processes are manually conducted, where check-in and baggage check are handled at the same counter. For passengers of business class and first class, and those requiring accessibility assistance, only traditional counters should be considered. In small to medium-sized regional airports, passengers usually have limited air travel experience and are less likely to adopt self-service check-in. According to Airport Development Reference Manual issued by IATA, among economy-class passengers, only 30% opt for self-service check-in, 5% opt for online check-in, and the remaining 65% prefer traditional check-in and ticketing (including traditional baggage check).

According to data from IATA, it is estimated that the number of passengers adopting self-service check-in is about 35% of the total passenger volume; about 25% of all passengers (70% of self-service check-in passengers) do not have checked baggage, and about 10% of all passengers (30% of self-service check-in passengers) need to check baggage.

Based on the given information that a one-and-a-half-story terminal typically has a throughput of no more than 2 million person-times, we can derive the passenger throughput that the given number of counters can satisfy, and some examples are shown in Tables 3-4 and 3-5 (counters are typically distributed in groups, so we round up to the nearest pair):

Table 3-4 Calculated and Suggested Number of Check-in Counters for Terminal Buldings of Different Sizes

Annual Passenger Throughput (ten thousand person-times)	Counters for Upper Classes and Accessibility Service	Traditional Check-in Counters	Self-service Baggage Check Counters	Total Number of Counters/Kiosks
≤ 95	2 (one set)	2	2 (one set)	6 (minimum number of counters)
100 ~ 130	2 (one set)	4	2 (one set)	8 (calculated number: 6.13 ~ 7.97)
135 ~ 160	2 (one set)	6	2 (one set)	10 (calculated number: 8.28 ~ 9.81)
165 ~ 195	2 (one set)	8	2 (one set)	12 (calculated number: 10.12 ~ 11.96)
200 ~ 225	2 (one set)	10	2 (one set)	14 (calculated number: 12.26 ~ 13.80)

Notes:
① Due to different calculation methods used by various design teams, there may be errors in the calculated results.
② "Total Number of Counters/kiosks" does not include check-in facilities in the VIP area (if there is any).

Table 3-5 Self-service Check-in Kiosks for Terminal Buildings of Different Sizes

Annual Passenger Throughput (ten thousand person-times)	Number of Self-service Check-in Kiosks	Annual Passenger Throughput (ten thousand person-times)	Number of Self-service Check-in Kiosks
≤ 50	2	140 ~ 160	6
55 ~ 80	3	165 ~ 190	7
85 ~ 150	4	195 ~ 215	8
110 ~ 135	5		

3.3.2.3 Spatial Arrangement of Check-in Ticketing Equipment

Traditional check-in and baggage handling facilities are typically grouped in sets, with each set comprising two counters, one dual-channel baggage security screening machine, and two baggage conveyor belts. Self-service check-in and baggage handling facilities also follow this pattern, with each set including two counters, one dual-channel self-service baggage security screening machine, and two baggage conveyor belts. In terminal design, except for oversized baggage handling equipment, these facilities are usually arranged in groups. Accessibility service counters are generally placed on one side of the check-in and ticketing area, close to the security check entrance.

For one-and-a-half-story terminals with lower annual passenger throughput, the number of check-in counters usually does not exceed 8 sets (16 counters). Half of these counters (about 4 sets) are arranged according to a pillar span of 9 meters or 10.5 meters, and the depth of the area for these counters should be no less than 8,500 millimeters plus the pillar diameter. The layout of the check-in counters is illustrated in Figure 3-25.

Self-service baggage check-in equipment is installed to meet the requirements for "safe, green, smart, and harmonious" airports. Currently, there are two main types of self-service baggage check-in equipment: modified traditional equipment and integrated self-service equipment.

Modified traditional equipment: During airport renovations, self-service baggage check-in modules can be added to existing traditional check-in equipment. This approach quickly increases the number of self-service baggage check-in facilities, catering to passengers on short trips. (Fig. 3-26)

Integrated self-service equipment: Newly built airports typically opt for integrated self-service baggage check-in equipment (Fig. 3-27). These units are highly integrated, reducing the risk of operation

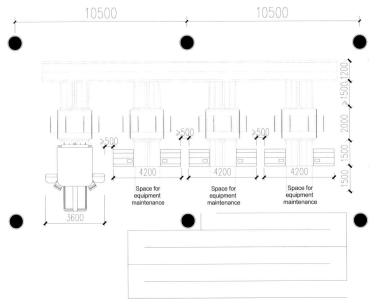

Figure 3-25 Layout of traditional check-in counters (including queueing space and 1-meter line, in millimeters)

Figure 3-26 Self-service baggage check-in modules in Terminal 2, Shanghai Hongqiao International Airport

Figure 3-27 Integrated self-service baggage check-in equipment at an airport in Xizang Autonomous Region

faults. Additionally, the height of the front end of such equipment from the ground is no more than 150 millimeters, making it convenient for passengers to place their baggage onto the conveyor belt, thereby enhancing check-in efficiency.

In addition to these two types of equipment, there is also oversized baggage check-in equipment specifically for the security screening of oversized baggage. This equipment is typically located on one side of the check-in area in the terminal and is directly connected to the baggage handling room.

3.3.3   Design Outlines of Ticketing Waiting Area

The ticketing waiting area is located in front of the check-in and ticketing area. Its design mainly considers the ways passengers queue for ticketing services and the demands for other functions.

3.3.3.1   Ways of Queueing for Check-in

In the ticketing waiting area, a 1-meter line serves as a divider: within this line is the ticketing area in front of the counters, and outside the line is the queueing and waiting area, typically designated by soft partition barriers. The queueing methods mainly include centralized queueing and queueing directly at the counters.

Centralized Queueing: All passengers first queue in a common space and are then directed to each counter for the ticketing process. This method is suitable for airports with high passenger throughput, especially when the depth of the check-in hall is shallow or when there is a high flight concentration. When deploying soft partitions, it is important to consider the spacing and depth between partitions, which helps in calculating the queue length. (Fig. 3-28)

Queueing at Counter: For regional airport terminals with deeper depth, queueing directly at counters can be adopted, although this may not be space-efficient. The queue length for each counter is generally considered to be no less than 15 meters, and soft partitions are used to longitudinally divide the ticketing waiting area.

3.3.3.2   Other Functions of Ticketing Waiting Area

Other important functions within the ticketing area include identity verification, identity document processing, and baggage unpacking checks.

Figure 3-28 Layout of 1.5-meter-wide soft partitions in the ticketing waiting area

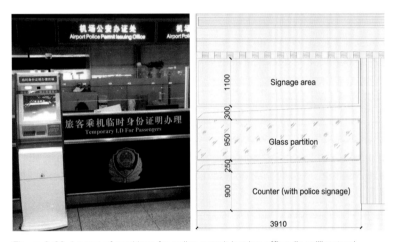

Figure 3-29 Layout of partitions for police permit issuing office (in millimeters)

1. Police Permit Issuing Office

The Police permit issuing office within the terminal primarily serves passengers without identity documents. Although the Civil Aviation Administration has released a mobile application for temporary document processing, considering the convenience of processing and special demands of passengers, it is still necessary to set up an open counter for manually permit issuing (with a length of no less than 2 meters). In smaller airports, this office can be combined with the police duty room. (Fig. 3-29)

2. Baggage Inspection Room

Used for inspecting checked baggage that does not comply with regulations, the baggage inspection room typically includes unpacking tables, security screening machines, and explosion-proof containers. There are two specific methods for unpacking and inspecting baggage: offline unpacking and

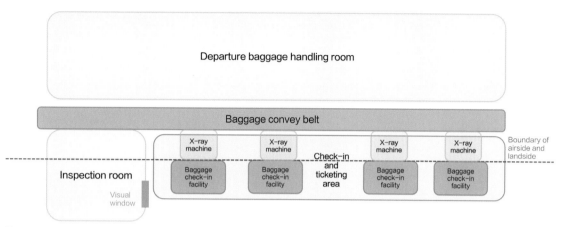

Figure 3-30 Independent inspection, centralized offline unpacking

island end unpacking.

Offline Unpacking: Suitable for check-in counters and facilities with X-ray inspection functionality that can identify non-compliant baggage directly during the ticketing process. Passengers then bring the baggage to the inspection room for further inspection and carry it back to the original ticketing counter for a second check-in after passing the inspection. This method is suitable for airports with fewer flights and lower passenger throughput. (Fig. 3-30)

Island End Unpacking: Used at check-in counters/equipment without X-ray inspection functionality. Checked baggage is directed through a diverter, and is conveyed to the X-ray machines for inspection, and passengers wait at the ticketing counter until their baggage passes the inspection. If the baggage does not pass inspection, they are required to enter the unpacking room for further inspection and complete the baggage check-in process in the same room if the baggage passes the inspection, without returning to the ticketing counter. This method is suitable for airports with a higher number of flights and passengers, where stricter baggage inspection is required (e.g. border airports), and demands more space for the unpacking room. It requires only two X-ray machines: one installed with the baggage carousel and the other placed within the unpacking room. (Fig. 3-31)

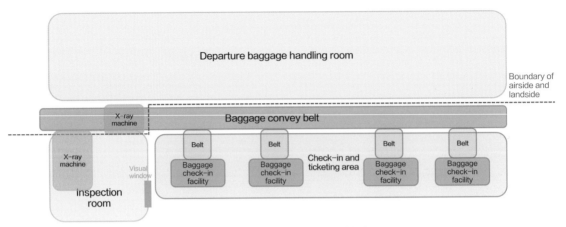

Figure 3-31 Centralized inspection, centralized unpacking (island end unpacking)

3.4 Design Outlines of Security Checkpoint/ Joint Inspection Area

3.4.1 Design Outlines for Different Types of Security Check Passages

  Security checks are a crucial component of the terminal process flow, serving as pivotal points between the airside and landside of the terminal. Therefore, the layout of security check passages plays a significant role in the construction of terminal security infrastructure. There are typically four types of security check passages:

  Passenger Security Check Passage: These passages are typically situated in prominent locations within the terminal, serving as essential nodes for departing passengers. To save passengers' time, these passages often connect to the departure hall and the boarding area. Passengers without checked baggage who have checked in online can proceed directly to the security check process upon arrival at the terminal.

  Two-Class VIP Security Check Passage: These can be positioned adjacent to the regular passenger security check passages for unified management and operation. If there is a separate VIP boarding lounge within the terminal with independent access, an additional VIP security check passage is required within the VIP boarding lounge, along with the corresponding staff and equipment.

  Return/Staff Passage: In airports with high flight concentration rates, the return/staff passage can be independently positioned alongside the regular passenger security check passages, parallel to the economy class passages. In airports with lower concentration rates, users of these passages may share the existing security check passages by scheduling their use during specific time periods to achieve user diversion.

Cargo Transport Passage: Cargo transport can share the staff passage, and it is advisable to be set up a separate passage if there are many commercial services in the terminal. Large floor-standing X-ray machines can be adopted to facilitate the transportation of goods and the removal of garbage. In smaller terminals, goods arrangement for commercials and garbage removal can be conducted during specific time periods, entering or exiting the airside via an access of the terminal area without a dedicated cargo passage.

3.4.1.1 Design Outlines of Passenger Security Check Passage

The placement of passenger security check passages in the terminal hall is critical, as it can determine the spatial layout of the entire terminal building. With the increasing imperative to ensure aviation safety, the update of passenger security check equipment has become a crucial factor in determining the size of security check passages.

The passenger security check passage primarily consists of a verification counter, queueing area, and check area. The location of baggage and passenger security check equipment determines the division between airside and landside areas within the terminal. Moreover, passengers spend the most time in this area during the departure process. Except for the queueing area, the width of a single check passage is recommended to be no less than 5 meters. If the diameter of columns within the passage is significant, it is suggested to increase the width. The depth of the check area typically ranges from 15 to 20 meters, depending on the type of equipment. Shallow depth may lead to insufficient space for equipment, resulting in congestion at the ends of the passage. (Fig. 3-32)

1. Verification Counter

After checking in and before entering queueing area for security check, passengers should have their tickets and carry-on baggage inspected to prevent those without tickets or those with oversized carry-on baggage from proceeding, which can lead to time waste during security checks. Ticket verification and oversized carry-on baggage checks can be performed manually or through self-service devices (common in large airports). Passengers who fail to pass the verification are not allowed to enter the queueing area.

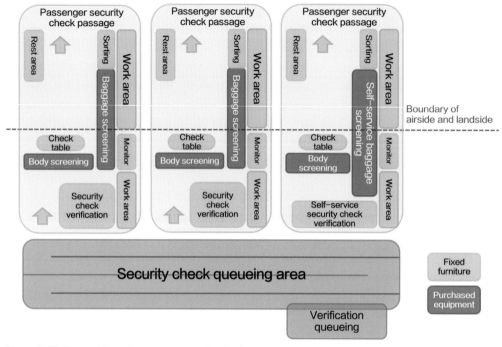

Figure 3-32 Composition of passenger security check passages

2. Security Check Queueing Area

There are two ways of queueing: centralized queueing and decentralized queueing (queueing in front of each check passage) (Fig. 3-33). Centralized queueing can enhance real-time operational efficiency, suitable for terminals with high flight concentration rates. Decentralized queueing is commonly used in medium to small airports, with queue lengths typically ranging from 15 to 20 meters. Longer queue lengths significantly decrease check efficiency.

At the ends of the security screening queueing area, there are usually prominent signs or display screens to alert passengers about items strictly prohibited on the aircraft. Trash bins are also provided for passengers to dispose of prohibited items they may have accidentally brought with them.

3. Verification Counter

The first step in the passenger/carry-on baggage security check is to vertify ticket and facial information. Passengers need to present valid identification documents (or valid identification provided by public security authorities) to verify their identity and flight information. There are two types of verification counters available: traditional manual verification and self-service verification.

4. Baggage/Passenger Security Checkpoint

To ensure aviation safety, the Civil Aviation Administration has strict regulations regarding carry-on baggage allowed on the aircraft. Passengers and their carry-on baggage must undergo thorough inspections in this area. Depending on the security equipment used, the time and steps of the check process may vary slightly. Airports with higher throughput and flight concentration rates tend to have more advanced equipment with higher inspection accuracy and fewer manual rechecks, which can significantly improve check efficiency. (Fig. 3-34)

5. Baggage Sorting Area/Passenger Rest Area

The item sorting table located behind the security check machine is intended for organizing baggage. Carry-on baggage passing the check can be directly conveyed to this area. Near the sorting table, there should be explosion-proof containers (which can be shared by two security check passages) in case of any hazardous materials such as flammable or explosive substances in carry-on baggage. Additionally, waste bins should be provided. If any prohibited items are found, they can be disposed of by staff after obtaining the passenger's consent. The seating area provided beyond the security check passages is for passengers to rest and organize dressing and other personal things.

Figure 3-33 Passenger security check queueing area in Terminal 2, Shanghai Hongqiao International Airport

Figure 3-34 Passenger security check area and staff in Terminal 2, Shanghai Hongqiao International Airport

3.4.1.2 Design Outlines of Return/Staff Passage

There are several types of staff, including airport staff, dispatched staff (working in leased areas), and airline staff. The inspection of flight crew members and their carry-on items is typically conducted at the main security check passages and can be shared with the return passengers from the airside boarding lounge. Designing staff security check passages involves two points:

1. Identity Verification

First, it is necessary to inspect the staff pass for airport control zone or the pass issued by the civil aviation administrative authority. Similar to passenger verification, staff identity verification requires facial recognition and camera imaging.

2. Carry-on Item Verification

All the belongings, goods, tools, small items, or devices of staff entering the boarding isolation zone must pass through the baggage security check machine. If there are oversized equipment or facilities, they are recommended to enter the airside of terminal through an airside passage or cargo station after security check.

3.4.1.3 Design Outlines of Cargo Transport Passage

For regional airports with one-and-a-half-story terminal buildings, the cargo passages can be combined with other security check passages. When the terminal area is less than 20,000 square meters and the leasable commercial area is large, it is recommended to combine security check passages for cargo and staff and set different time periods for utilization, respectively. Large cargo can be transported through specialized access or the flow line of the cargo station. When the terminal area exceeds 20,000 square meters, cargo passages can be independently set up and share stretcher elevators (or fire elevators) to improve the efficiency of cargo transportation and garbage removal within the terminal.

3.4.2 Selection of Passenger Security Check Equipment

The passenger security check passage is crucial for ensuring aviation safety. It provides strong protection for passengers' property and lives, and its efficiency also determines that of the terminal's functional processes. Therefore, the equipment in the security check area of the terminal requires frequent

Figure 3-35 Equipment layout of passenger security check passages in Terminal 2, Shanghai Pudong International Airport

Figure 3-36 Manual security verification counter

upgrades, and different airports choose different equipment. (Fig. 3-35)

3.4.2.1 Security Verification Counter

The most critical step for passengers on the landside before boarding is to verify tickets and passenger information at the security verification counter. The verification process is interconnected with civil aviation weak power systems and public security systems. There are mainly two types of security verification counter:

1. Traditional Security Verification Counter

The traditional security verification counter is a type of fixed furniture inside the terminal building, equipped with civil aviation weak power interfaces to meet the requirement for both camera and manual identification of passenger identity simultaneously. Each security check passage requires one verification counter and one staff member. Since 2020, for the requirements of globally standardized epidemic prevention, protective isolation screens are required for all passenger security verification counters (Fig. 3-36). For exclusive security passages of VIP lounges, traditional security verification counters are mostly chosen to enhance service quality and efficiency.

2. Self-Service Security Verification Counter

To meet China's appeal for smart airport construction and improve the reliability of security verification, most terminal buildings in China now advocate for fully self-service airport process implementation. Thus, self-service security verification counters have emerged, which only require passengers to present their second-generation ID cards, providing smooth and convenient verification. However, for passengers who cannot handle the self-service, manual verification windows alongside self-service verification counters are still necessary to help them. (Fig. 3-37)

Figure 3-37 Self-service security verification counter in a certain airport

3.4.2.2 Selection of Security Check Equipment

Passenger security check includes baggage inspection and passenger body inspection. Baggage inspection primarily involves screening for items prohibited by the Civil Aviation Administration from being brought onto the aircraft. Passenger body inspection primarily focuses on detecting prohibited items (mainly metal products) carried by passengers.

1. Baggage Inspection Equipment

Baggage inspection equipment is primarily categorized into two types based on working principles: X-ray machines and CT machines. Based on the operation mode of the baggage basket, they are classified into traditional manual sorting machines and those with a basket recycling system. Three staff members are required beside the baggage screening equipment, one for baggage preparation, one for reviewing the screening images (baggage interpreter), and one for baggage unpacking.

Regardless of the equipment type, the determination of the baggage safe or not mainly relies on baggage interpreters. They assess the types and sizes of items in the baggage according to the regulations of the Civil Aviation Administration and conduct secondary inspections of items that may not comply with the regulations.

X-ray baggage screening equipment: X-ray machines are the most widely used and extensively deployed baggage screening equipment in Chinese airports. These machines, mature among common security screening equipment types, are relatively small in size and cheaper compared to other types. The mechanism of X-ray inspection is projection photography, by which items inside baggage are presented as overlapping two-dimensional images. To decrease the interpretation pressure, complex electronic items

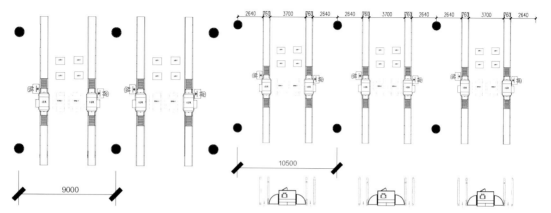

Figure 3-38 Layouts of X-ray security screening machines for different pillar spans (in millimeters)

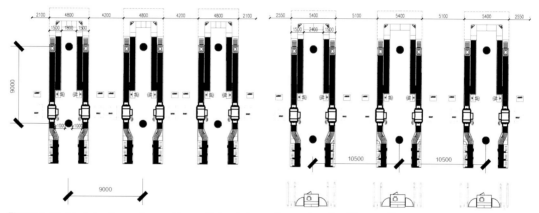

Figure 3-39 Layouts of security machines with basket recycling system for different pillar spans (in millimeters)

need to be taken away before inspection. Larger airports have higher requirements for the skill of baggage interpreters. (Fig. 3-38)

Screening equipment with basket recycling system: Both CT and X-ray screening machines can be equipped with a basket recycling system. This system provides automatic circulation of the baggage baskets during baggage inspection. Passengers can retrieve a basket at the front end of the security screening equipment, place their baggage in it for security check, and then the baskets can be automatically recycled and conveyed back to the front end of the equipment. This process significantly enhances check efficiency and reduces labor costs, especially in airports with a high flight concentration rate. With the addition of the basket recycling system, the depth of the security equipment area should be further extended (by approximately 3 to 5 meters) to accommodate the recycling system. (Fig. 3-39)

2. Passenger Body Screening Equipment

In addition to ensuring the security of baggage, conducting security checks on passengers themselves is equally important. This not only enhances aviation safety but also aims to minimize illicit carry-on items aboard. Passenger body screening equipment typically includes metal detection gates and millimeter wave security gates.

① Metal Detection Gate

This type of detection gate has been used in Chinese airports for up to 30 years, primarily employing weak magnetic field technology. It detects and alarms metal objects carried by the human body, without missing or mistakenly reporting them. By adjusting the sensitivity of the detection gate, the size of alarmed items can be managed. Generally, the sensitivity of security detection gates at airport passenger security check passages is high enough to prevent prohibited items. Therefore, security check passages equipped with such gates typically adopt a combined body inspection mode of "metal detection by security gates + manual comprehensive body search." (Fig. 3-40)

② Millimeter Wave Security Gate

The millimeter wave security gate represents an advanced type of security equipment that balances security effectiveness with human safety. Its imaging mechanism involves emitting millimeter waves from various surfaces of the device towards passengers. Due to differences in reflectivity between human skin, clothing, and other materials, the device can scan and detect areas that differ from normal patterns. If a passenger is carrying prohibited items concealed within their clothing (items hidden inside the body cannot be detected), the gate will display clear surface and outline images of these items for security personnel to interpret.

This technology enables rapid detection of suspicious items, regardless of whether they are metallic, non-metallic, solid, liquid, or even

Figure 3-40 Metal detection gate

packaged hazardous gases. It significantly reduces the workload of security personnel, improves the capability to inspect prohibited items, and eliminates the need for physical contact with passengers during the inspection process. Furthermore, passengers are not required to remove clothing or turn around in place during the inspection, enhancing the overall comfort of the security screening process.

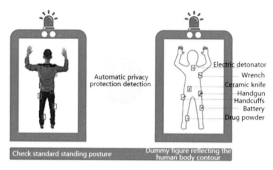

Figure 3-41 Inspection mechanism of millimeter wave security gate

The millimeter wave security gate operates on a mode of "comprehensive inspection+automatic alarm-assisted body search." Each check passage equipped with such a device can save 1 to 2 manual body searchers, and the entire process typically takes only 2 seconds per passenger. However, the cost of deploying millimeter wave security gates is higher, which may require careful consideration, especially for small- and medium-sized airports balancing cost-control factors. (Fig. 3-41)

3.4.2.3 Arrangement of Sorting Table and Rest Area

Sorting tables and seating for rest areas are typically positioned at the end of the baggage inspection equipment. In terms of design, the passenger side should facilitate passengers in organizing their belongings after baggage inspection, while the staff side should provide space for storing items. A typical design of the sorting table is illustrated in Figure 3-42.

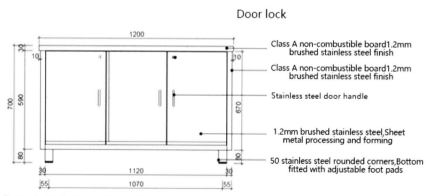

Figure 3-42 Sorting table design drawing (in millimeters)

3.4.2.4 Security Partition

Security partitions are generally positioned behind the security verification counter, serving as physical barriers between the airside and landside areas. Whether traditional or self-service verification counters are used, these partitions can be solid walls or glass barriers, with a height requirement of no less than 2.5 meters and designed without footholds (crossbars) to prevent climbing, meeting security standards.

The partitions between security check passages and those behind them can be configured as per the airport security personnel's requirements. It is crucial to prevent any direct sightlines that could allow oversight of the entire security check area from either the front or the back of the security check passage.

3.4.3 Calculation of Security Equipment Quantity

For a one-and-a-half-story terminal building, the security checkpoint/joint inspection area serves as a crucial hub connecting the terminal's airside and landside. The number of security equipment and facilities significantly impacts the width and efficiency of operational processes in this connecting area.

According to IATA guidelines, calculations are based on the throughput of 10-minute peak periods at ticket counters and the number of passengers in queues to determine if the security verification counters meet design standards. Assuming a security check efficiency of 120 people per hour (30 seconds per person) and a maximum waiting time of 10 minutes, airports with an annual passenger throughput typically not exceeding 2 million passengers are categorized into intervals of 500,000 passengers to calculate the required security check equipment quantities, as detailed in Table 3-6.

Table 3-6 Quantities of security check equipment required by regional airport terminal building in different sizes

Annual Passenger Throughput (ten thousand person-times)	Traditional Verification Counters	Self-service Verification Kiosks	Security Check Passages
50	4 (calculated result is 3.5)	⩾ 1	4
100	6 (calculated result is 5.02)	⩾ 1	6
150	7 (calculated result is 6.54)	⩾ 1	7
200	9 (calculated result is 8.05)	⩾ 1	9

3.5 Design Outlines of Boarding Lounge

3.5.1 Layout Outlines of Boarding Lounge

The spatial layout of the boarding lounge mainly includes spatial capacity calculation and column grid layout design. The former involves the calculation of the number of seats needed and the area required for the boarding lounge.

3.5.1.1 Spatial Capacity Calculation

The number of seats needed can be deduced based on the number of seats on the aircraft (i.e., the passenger capacity of the aircraft). Since small regional airports generally serve aircraft of 4C to 6C levels, the calculation of seat quantity is mainly based on C-Level aircraft.

The seat quantity needed in the boarding lounge = n × Aircraft passenger capacity × 80% (shared coefficient for the boarding lounge) × 80% (seated passengers in the boarding lounge)

Where n can be set at 80% according to IATA standards, but Chinese experts suggest it should increase to 90% as the number of passengers has increased in recent years.

Then the floor area of the boarding lounge can be calculated.

Based on IATA standards, each seated passenger requires at least 1.7 square meters, while each standing passenger requires 1.2 square meters. Thus, the required area of boarding lounge is calculated as follows:

Required area of boarding lounge = [(80% × Aircraft passenger capacity × 80% × 1.7) + (80% × Aircraft passenger capacity × 20% × 1.2)] ÷ 65% (maximum occupancy rate of seating area for a Class-C aircraft in the boarding lounge)

For a Class-C aircraft with a full passenger capacity of 180, the calculations for a regional airport are as follows:

Seat quantity needed in boarding lounge = 0.8 × 180 × 0.8 × 0.8 ≈ 92;

Required area of boarding lounge = [(0.8 × 180 × 0.8 × 1.7)+(0.8 × 180 × 0.2 × 1.2)] ÷ 65% ≈ 350 (square meters)

Additionally, different boarding methods impact the ratio of standing to seated passengers in the boarding lounge, and this factor should be considered when calculating the number of seats needed. For example, if the near parking stand boarding mode is mainly used, for a Class-C aircraft, the number of people requiring seats in a boarding lounge is about 92, requiring a total boarding lounge area of about

350 square meters according to the formulas above. However, due to the lower concentration of remote stand gates and fewer remote stand gates (generally no more than 3) in small regional airports, a higher shared coefficient can be considered for seats in their boarding lounges. Thus, the number of fixed seats for remote stand gates can be fewer than the number for near parking stand gates.

3.5.1.2 Scale of Column Grid and Seat Arrangement

Based on the optimal solution for the airside apron line, the most economical spacing of the column grid is 9 meters and 10.5 meters. Using this spacing, the layout of the column grid and seat arrangement can be deduced as follows:

With a column grid spacing of 9 meter, a minimum spacing between rows of seats as 3 meters, and a seating area width of 5 meters, a boarding lounge of 350 square meters can accommodate 108 seats. (Fig. 3-43)

With a column grid spacing of 10.5 meters, a spacing between rows of seats of 3.3 meters, and a seating area width of 5 meters, a boarding lounge of 350 square meters can accommodate approximately 120 seats. (Fig.3-44)

Besides, as the width of the standing area can be narrowed to a certain degree, the depth of the commercial area can be increased accordingly under the premise of a fixed total depth of the commercial area and the boarding lounge.

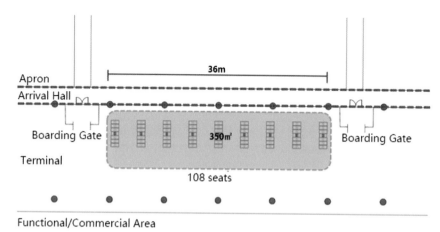

Figure 3-43 Layout of boarding lounge with 9-meter-spacing column grid

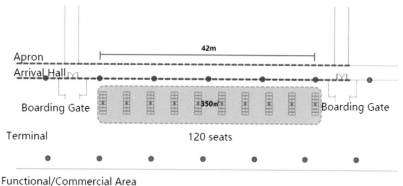

Figure 3-44 Layout of boarding lounge with 10.5-meter-spacing column grid

3.5.2 Types of Seats for Boarding Lounge

The seats in boarding lounges can generally be categorized into three types based on their functionality: regular seats, business leisure seats, and creative leisure seats.

Regular seats are intended for departing passengers to rest inside the terminal building. The number of regular seats can be calculated using the methods mentioned above. These seats are arranged in rows for passengers' comfort and reasonable utilization, with a recommended spacing between rows of 3 to 3.6 meters. (Fig. 3-45)

Business leisure seats are relatively flexible in arrangement and are usually equipped with power outlets and tables for office use. (Fig. 3-46)

Figure 3-45 Layout of regular seats in boarding lounge

Figure 3-46 Conceptual design of business leisure seats

Figure 3-47 Conceptual design of creative leisure seat area

Creative leisure seats have more flexibility and variety in arrangement. They can be combined with leisure businesses, themed events, and other activities within the terminal building to create distinctive leisure spaces. (Fig. 3-47)

3.6   Design Outlines of Boarding Gate

3.6.1   Spatial Dimensions of Boarding Gate

There are two types of boarding gates: self-service boarding gates and traditional manual check boarding gates.

A self-service boarding gate generally consists of flight display devices, self-service ticket check machines, manual ticket check counters, boarding gate signage, surveillance equipment, etc. It usually has a width of approximately 5 meters and a depth of around 4 meters. The width of the self-service check

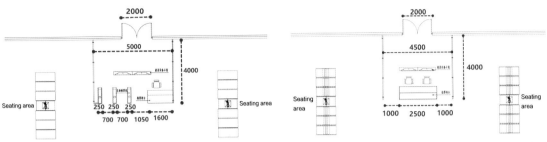

Figure 3-48 Spatial dimension schematic of self-service boarding gate (in millimeters)

Figure 3-49 Spatial dimension schematic of manual check boarding gate

passage is about 0.75 meters, while the manual check passage is about 1.05 meters. (Fig. 3-48)

The device composition of manual check boarding gates is relatively simple, typically including flight display devices, manual ticket check counters, boarding gate signage, etc. A manual check boarding gate has a width of approximately 4.5 meters, a depth of around 4 meters, and a check passage width of about 1 meter. (Fig. 3-49)

3.6.2 Layout of Boarding Gate for Near Parking Stand

The boarding gate for near parking stand is connected to the jet bridges via the arrival corridor. The layout of such boarding gates is mainly determined by the boarding methods for passengers and should also consider the relationship with the arrival corridor.

Boarding Methods for Passengers: Due to differences in the principles of layout between airside jet bridges and terminal boarding gates, as well as various control factors between jet bridges and boarding gates, several points need consideration: ① Whether the layout can meet the requirements of security access control; ② The boarding lounge, arrangement of gates, the flow of service vehicles, and other facilities and equipment all affect the location of the boarding gates; ③ Whether the passenger arrival flow line adopts a seperated or mixed mode. (Fig. 3-50)

Layout of Queueing Area for Boarding: The queueing area of the boarding gate is usually positioned directly in front of the boarding gate. In special circumstances, the queueing area can also be arranged on both sides of the boarding gate, parallel to the arrival corridor. (Fig.3-51)

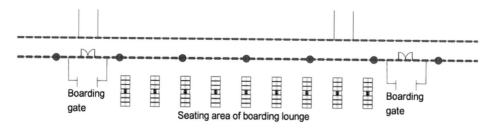

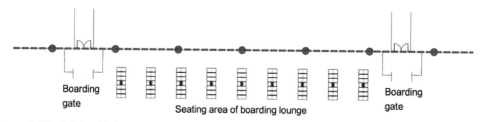

Figure 3-50　Relationship between boarding gate and jet bridge

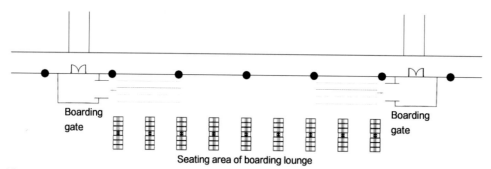

Figure 3-51　Relationship between boarding gate and arrival corridor

3.6.3 Boarding Gate for Remote Parking Stand

Apart from boarding gates for near parking stands close to the terminal building, boarding gates for remote stands are also required, which passengers reach by shuttle bus.

When positioning boarding gates for remote stands, architectural elements such as roof eaves or terrace on the second floor can be utilized as awning for shuttle buses. (Fig. 3-52)

In northern regions of China with extremely cold weather in winter, the porch of the remote boarding gate can be positioned inside the terminal building, making it safer and more comfortable for passengers to board the shuttle bus. (Fig. 3-53)

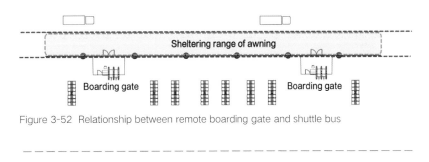

Figure 3-52 Relationship between remote boarding gate and shuttle bus

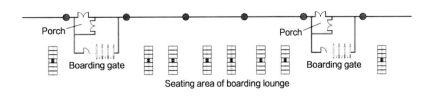

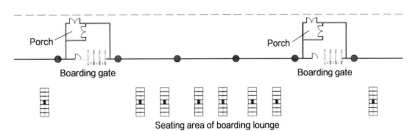

Figure 3-53 Relationship schematic of the porch of remote boarding gate

3.7 Design Outlines of Baggage Claim Hall

3.7.1 Spatial Dimensions of Baggage Claim Hall

After arriving at the terminal building, passengers first retrieve their baggage in the baggage claim hall, then conduct baggage inspection and ticket checks to enter the public hall.

3.7.1.1 Baggage Carousel Scale Analysis

Baggage carousels can typically be classified into flat carousels and inclined carousels based on the panel type (Fig. 3-54). Based on the position relationship between the baggage handling room and the baggage claim hall, baggage carousels are classified into same-level carousels and multi-level carousels. The former is commonly used in small airports, while the latter is typical in large airports. Based on the planar shape of the carousel, they can be classified into straight, L-shaped, T-shaped, U-shaped, etc. Small airports often utilize T-shaped carousels. Considering the length of the baggage trailer (approximately 13 meters), the distance between two baggage carousels is typically about 10 to 13 meters. (Figs. 3-55, 3-56)

Figure 3-54 Flat carousel (left) and inclined carousel (right)

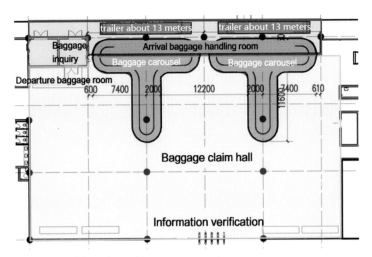

Figure 3-55　Dimensions of baggage carousels

Figure 3-56　Baggage carousel in Frankfurt Airport, Germany

3.7.1.2 Scale of Near-Gate Arrival Area

Considering the connection of the baggage claim hall with the passenger arrival flow, for those arriving via the near gate, as the baggage claim hall is typically located on the ground floor, passengers arriving at near gates generally descend from the second floor to the ground floor via escalators (1.6-meter-wide), elevators (2.8-meter-wide), and stairs (1.6-meter-wide). The width of the entrance for these pathways is typically around 6 meters, accommodating 3 to 4 streams of parallel pedestrian traffic. (Fig. 3-57)

3.7.1.3 Scale of Remote-Gate Arrival Area

For arrivals at remote gates, passengers typically disembark the aircraft and take airport shuttle buses to the ground floor arrival gate for the remote gates before entering the baggage claim hall. Considering the flow of passengers carrying baggage such as rolling suitcases, with an average movement width of about 3.4 meters, it is advisable to widen the width of entrance for remote-gate arrival to around 4 meters, accommodating 3 to 4 streams of traffic comfortably. (Fig. 3-58)

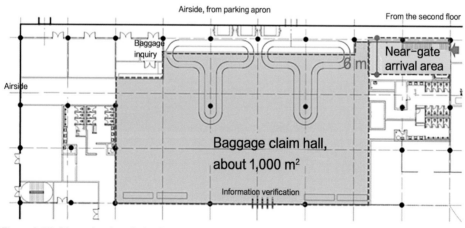

Figure 3-57 Dimensional analysis of near-gate arrival area

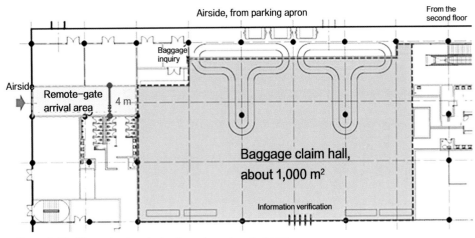

Figure 3-58 Dimensional analysis of remote-gate arrival area

3.7.1.4 Space for Oversized Baggage Retrieval and Baggage Inquiry

The oversized baggage retrieval area typically includes an oversized baggage inquiry zone, an oversized baggage storage zone, and service counters. Once oversized baggage arrives at the arrival baggage handling room, it would be delivered to passengers at service counters via baggage inquiry. Additionally, express delivery services are often provided as well. (Fig. 3-59)

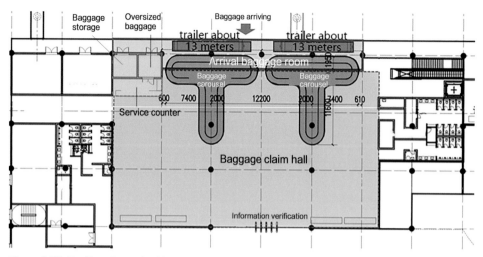

Figure 3-59 Position of oversized baggage retrieval

3.7.2　Arrival Baggage Verification Process

After passengers retrieve their baggage from the baggage carousel in the baggage claim hall, they proceed to the exit to verify whether the baggage belongs to them. Upon successful verification, they can proceed to the exit hall.

Baggage Tag Verification: Typically, at the exit of the baggage claim hall, staff members check the boarding passes and the barcodes on the baggage using scanners or manually to ensure they match. Once the verification is passed, passengers are allowed to proceed. (Fig. 3-60)

Other Information Verification: In certain special circumstances, airports may implement additional verification procedures such as identity verification, infectious disease checks, etc. These checks typically occur simultaneously at the document inspection points, which may lead to significant congestion due to increased passenger flow. To address this, airports may increase the number of inspection passages. (Fig. 3-61)

Figure 3-60 Tag verification on arrival baggage

Figure 3-61 COVID-19 testing passage set up at Shanghai Pudong International Airport during pandemic

3.8 Design Outlines of Baggage Handling Area

3.8.1 Baggage Handling Room

The baggage flow line is one of the most critical flow lines within the terminal, with the primary requirement for baggage handling area being simplicity, efficiency, and facilitating smooth movement for baggage trailers to enhance loading and unloading efficiency. The baggage handling area primarily consists of the departure baggage room and the arrival baggage room, with six key design outlines: ① Minimize the stages of baggage transportation; ② Reduce the turns and spatial changes for baggage flow; ③ Ensure that the slope of baggage carousel does not exceed 15 degrees; ④ Place the baggage sorting area as close to the apron as possible; ⑤ Separate departure and arrival baggage rooms to avoid crossover of their flows; ⑥ Implement intelligent, paperless, environmentally friendly, and energy-efficient systems.

3.8.1.1 Departure Baggage Room

The departure baggage room should be located near the apron to receive passengers' baggage conveniently and transport it to the departing flights promptly via baggage transport vehicles. (Fig.3-62)

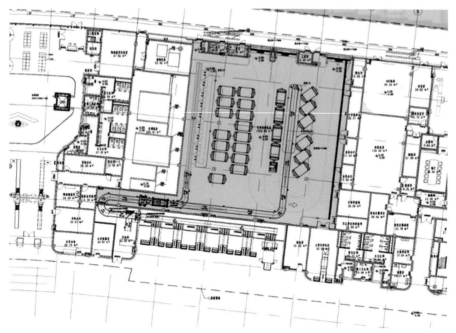

Figure 3-62 Layout of departure baggage room in a certain airport

For regional airports, the overall volume of the terminal building is relatively small, and the amount of baggage to handle is limited, suitable for manual baggage sorting. For airports with high flight concentration rate, automated sorting can also be considered.

Baggage is divided into standard baggage and oversized baggage based on size. Standard baggage departure pathway (manual sorting) is: Check-in counter → Baggage Security check → Departure conveyor belt → Departure carousel/chute. Standard baggage departure pathway (automated sorting) is: Check-in counter → Baggage Security check → Departure conveyor belt → Sorting machine → Departure carousel/chute. (Figs. 3-63, 3-64)

Figure 3-63 Automatic sorting system in departure baggage room of a certain airport

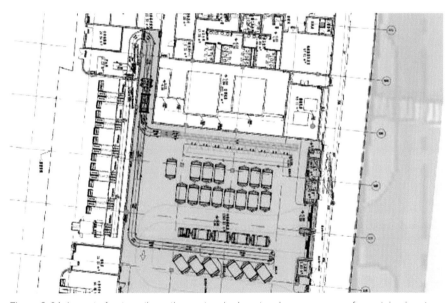

Figure 3-64 Layout of automatic sorting system in departure baggage room of a certain airport

Oversized baggage is handled directly at the oversized baggage counter for check-in, and is manually delivered to the baggage room.

3.8.1.2 Arrival Baggage Room

The arrival baggage room should also be adjacent to the apron, convenient for the reception of baggage delivered by aircraft through baggage transport vehicles and transportation to the arrival baggage handling area. (Fig. 3-65) After the aircraft arrives, baggage is transported to the unloading zone via baggage transport vehicles. The baggage is then placed on the conveyor belt and transported to the baggage claim hall through the carousel system. Typically, one transport vehicle can carry baggage for 80 to 100 passengers.

Figure 3-65 Layout of arrival baggage room

3.8.2　Baggage Handling System

The baggage handling system includes the departure baggage system, the arrival baggage system, and the oversized baggage system.

The main equipment and facilities for baggage handling include departure check-in lines, check-in counters (including standard baggage departure counters and oversized baggage departure counters), self-service baggage check-in counters, X-ray security screening machines (including dual-channel X-ray machines and large-channel X-ray machines), departure carousels, arrival carousels, and the control system.

3.8.2.1　Baggage Handling Processes

1. Departure Baggage System

The departure baggage system handles standard baggage from departing passengers. It consists of the check-in island, conveyor lines, and departure carousel. The journey starts at the check-in island counters and ends at the departure sorting carousel. (Fig. 3-66)

If the departure baggage room is small, only one side of the departure carousel may be available for trailers to load and unload baggage.

2. Arrival Baggage System

The arrival baggage system includes carousels installed between the arrival baggage unloading area and the baggage claim hall for domestic arriving passengers to retrieve their baggage. The arrival baggage is directly unloaded onto the retrieval carousel. Since baggage trailers enter the handling area, bollards should be installed between the carousel and the trailer lane to prevent collisions. (Fig. 3-67)

3. Oversized Baggage System

The oversized baggage system includes departure and arrival oversized baggage systems. The departure oversized baggage system is equipped with departure oversized baggage counters and large-channel X-ray security screening machines. Oversized baggage that passes security screening is manually transported to the loading area and uploaded onto trailers. The arrival oversized baggage is manually transported to the arrivals baggage claim hall.

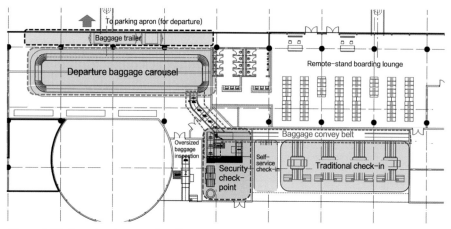

Figure 3-66 Departure baggage handling system

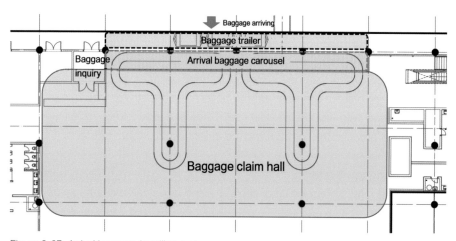

Figure 3-67 Arrival baggage handling system

3.8.2.2 Security Screening System for Checked Baggage

All checked baggage should undergo security checks, with many facilities currently utilizing multi-source dual-view X-ray security inspection equipment. Departing checked baggage mostly undergoes security checks using the "dual-channel X-ray security screening machine" at the counters. The system follows a hierarchical management model with on-site unpacking inspections.

Adjacent to the transport counters, there is typically a room for image interpretation and unpacking: image interpreters analyze images in this room; suspicious baggage is automatically returned, and security personnel take it to the unpacking area to inspect it together with the passenger, ensuring passenger privacy is protected as much as possible.

Baggage that passes the unpacking inspection must be returned by the passenger to the counter for transportation and undergo further inspection.

Departing oversized checked baggage undergoes inspection using large-channel X-ray security screening machines. Suspicious baggage undergoes on-site unpacking inspection (if the baggage size permits, it may also be taken to the unpacking area for inspection). (Fig. 3-68)

Figure 3-68 Baggage security check machine

3.8.2.3 Baggage Belt

The baggage belt is a conveying facility within the baggage system, typically 1.2 meters wide with a maintenance space of at least 600 millimeter reserved on both sides. Between the baggage room and the other public areas, there should be fireproof roller shutters or rubber curtain doors as partitions, with the minimum clear height of the door being no less than 1.6 meters. (Fig. 3-69)

3.8.2.4 Baggage Trailer

The operation flow of the baggage trailer is: firstly, to transport the arrival baggage from the flight to the arrival baggage handling room to unload onto the carousel; secondly, to proceed to the departure carousel to receive the departure baggage and transport them to corresponding flights. (Fig. 3-70)

A standard baggage trailer is 1.8 meters wide and occupies a lane of 2 meters in width, while a typical trailer lane is 3 meters wide. Each trailer section has a length of 3.8 meters, and one trailer can tow a maximum of 4 sections. Therefore, the length of a four-section trailer is 13 meters. (Fig. 3-71)

For a Class-C aircraft, the loading, unloading and transportation can be conducted at only one time by the trailer. The economical length of departure carousels used for baggage handling can be determined based on flight allocations and trailer length. Generally, the length for the straight segment of 20 meters is considered the most efficient.

Figure 3-69 Baggage conveyor belt

Figure 3-70 Baggage trailers

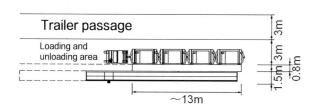 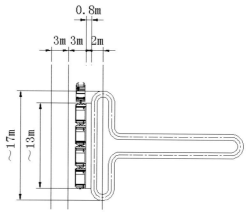

Figure 3-71 Spatial layout of baggage trailer at the departure baggage carousel

Figure 3-72 Spatial layout of baggage trailer at the arrival baggage carousel

The length of the unloading segment of one arrival carousel should be set according to the unloading requirements of the trailer, typically allowing 3 to 4 sections to unload at the same time, which requires at least 13 meters. (Fig. 3-72)

3.9 Design Outlines of VIP Lounge

3.9.1 Types of VIP Lounges

The VIP lounge in a terminal building is a critical, independent and significant node, usually positioned upstream of the main flow (vehicle circulation) and connected to the departure hall via designated passages. The VIP lounges are generally categorized into the following types:

Government VIP Lounge: Typically operated independently by the airport company or outsourced to third-party operators, designed to meet the needs of local government receptions. The level of the VIPs determines the standards for various reception tasks, functions, and security levels. The flow line for government VIPs does not intersect with the main flow; dedicated personnel handle processes like ticketing and baggage check-in, with separate security check passages in the VIP lounge.

Business VIP Lounge: Primarily for important business individuals whose processes may impact

Figure 3-73 Conceptual rendering of government and business VIP lounge of a certain airport

regular passengers. Usage frequency varies by airport location. In small- and medium-sized airports, this lounge may combine with the government VIP lounge, sharing similar services and processes (Fig. 3-73).

Cardholder and Two-Class VIP Lounge: Typically operated by airlines, but may be centrally operated by the airport company in smaller airports. Targeted at frequent business travelers who require convenient VIP lounge services but do not meet the standards for independent boarding in the VIP area.

These passengers share departure/arrival processes with economy-class passengers but have exclusive counters and passages for ticketing and security checks. The corresponding VIP service areas are usually located where waiting and boarding are convenient, primarily providing temporary rest space for business travelers. In airports with the necessary facilities, it is recommended to set up independent quick boarding passages.

3.9.2 Key Points of VIP Lounge Design

VIP lounges are categorized into landside and airside lounges based on the security checkpoint location. Landside lounges are primarily set up for government officials and business VIP passengers, while airside lounges are primarily for the cardholder and upper-class passengers.

3.9.2.1 Landside VIP Lounge

Landside VIP lounges have stricter privacy and security requirements and are mainly set up as private rooms, which may contain independent dining areas. If spacious enough, dining areas may be centralized within the lounge. Specific usage needs for different types of VIP passengers are discussed as follows. (Fig. 3-74)

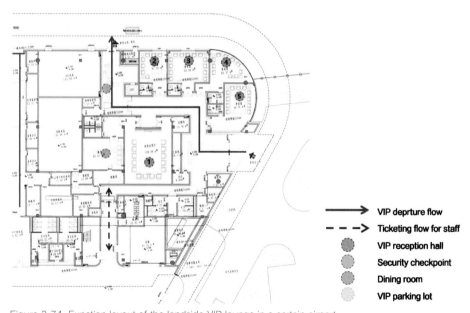

Figure 3-74 Function layout of the landside VIP lounge in a certain airport

1. Departing VIP Passenger Need

Departing VIP passengers tend to spend more time in the VIP lounge to wait for staff to handle check-in, ticketing, and baggage processes. Departing VIPs may rest briefly in the reception hall of the VIP area when the flight departure time is near, or stay longer in the VIP lounge if they have more time before departure. Therefore, the layout of the reception hall in the VIP lounge should accommodate the temporary usage needs of VIPs (traveling in groups of at least three)(Fig. 3-75).

To facilitate various services for departing VIPs, if separate check-in counters are not provided in the VIP area, there should be convenient access between the reception hall of the VIP lounge and the check-in and ticketing area of the terminal's departure hall, reducing waiting times for departing VIPs.

2. Arriving VIP Passenger Need

Landside VIP lounges should consider the convenience of arriving VIPs leaving the terminal building. Staff should retrieve their checked-in baggage and deliver it to the entrance/exit of the VIP lounge, equipped with a road system for convenient drop-off, circling, and parking for VIP passengers.

Given that arriving VIPs typically spend a short time in the terminal building, small public restrooms can be placed near the reception hall of the VIP lounge to meet temporary needs of VIP passengers.

3. Greeting/Sending-off Entourage of VIP Passengers

In addition to departing and arriving VIP passengers, landside VIP lounges should consider the needs of entourage greeting or seeing off these VIPs. Apart from parking areas, a separate area adjacent to the reception hall can be designated for these people. If space permits, it is advisable to provide a special rest area for longer stays. (Fig. 3-76)

Figure 3-75 Concept image of VIP reception hall in a certain airport

Figure 3-76 Layout of VIP rest area of a certain airport

3.9.2.2　Airside VIP Lounge

Airside VIP lounges inside the terminal primarily serve cardholder and upper-class VIP passengers, offering a smaller service radius. Therefore, remote/near parking stand boarding lounges and domestic/international boarding lounges should have airside VIP lounges. These lounges typically include tea seating area, individual rest area, private rooms, and service facility areas. (Fig. 3-77)

1. Tea Seating Area

The tea seating area in airside VIP lounges caters to short stays (less than 2 hours), allowing for temporary rest and tea tasting before boarding. To simplify the boarding process, staff can transport VIP passengers' carry-on baggage to their designated seats on the flight if provided conditions allow.

The tea area should accommodate small groups of 3 to 5 passengers and be equipped with restrooms, showers, and dining zone for light meals. (Fig. 3-78)

Figure 3-77　The V1 VIP lounge (at airside) of Shanghai Hongqiao International Airport

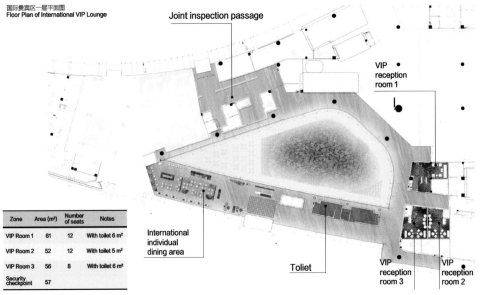

Figure 3-78 The tea seating area of airside VIP lounge in a certain airport

2. Individual Rest Area

The individual rest area for airside VIP lounges is designed for longer stays (2 to 3 hours). Seats are mostly arranged individually, providing private charging and entertainment facilities. This area is quiet and private, often located at the end of the VIP lounge or in the form of independent work cabins. (Fig. 3-79)

3. Private Room

Airside VIP lounges also have private rooms, which are relatively smaller than those in landside VIP lounges. They can be set up as medium-sized rooms for 10 people or small rooms for 3 to 5 people. Food services are usually not provided inside private rooms, but light meals can be provided if needed.

Figure 3-79 Concept image of VIP reception hall in the satellite hall of a certain airport

3.9.2.3 Service Facilities for VIP Lounge

The public area of landside VIP lounges is equipped with VIP offices, public service rooms (with separate service rooms inside the private rooms), meal preparation areas (if there is dining zone), duty rooms, and cleaning rooms. For cardholder and upper-class VIP lounges, it is common to have only duty rooms and comprehensive service rooms.

Each service room in a VIP lounge has an area of approximately 6 to 10 square meters. These rooms need to be equipped with independent water supply and drainage systems, refrigeration units, and workstations with computers for office work.

There are two methods for meal provision in VIP lounges: delivery and in-house preparation.

Delivery requires only a food preparation room in the dining zone. In-house preparation requires food storage room, transportation/unloading passages, kitchen, grease separation room, and food preparation room. If gas is required for food processing, the kitchen must be located adjacent to an external wall, not adjacent to crowded areas, and the gas-related facilities must be designed by qualified design institutes.

SPATIAL COMPOSITION

SPATIAL COMPOSITION OF
ONE-AND-A-HALF-STORY TERMINAL BUILDING

4.1 Types of Spatial Combination of One-and-a-Half-Story Terminal Building

The one-and-a-half-story terminal features a partial second floor, with both arrival and departure functions sharing the same level. Based on the different functions laid out on the second floor, it can be categorized into three combinations:

Combination Ⅰ : Boarding lounge Located in the Upper Part

Departure and arrival flow lines are closely connected. Almost all processes are arranged on the first floor of the terminal, with only the near-stand boarding lounge on the second floor.

Combination Ⅱ : Security Check and Boarding lounge Located in the Upper Part

The check-in area and arrival processes are on the first floor, while the security check and departure processes, including the boarding lounges, are on the second floor.

Combination Ⅲ : Same-Level Entry and Exit, Layered Departure and Arrival

The entry and exit are on the same level, but the processes are independent of each other. Arrival processes are on the first floor, while departure processes are on the second floor.

In the following sections, the characteristics, corresponding cases, advantages, and disadvantages of these three types of airport terminal spatial combinations are analyzed in detail, providing a systematic reference for the selection of regional airport terminal configurations.

Table 4-1: Three Types of Spatial Combination of Terminal Buildings

Type	Location	Functional Area						
		Departure Hall	Check-in and Ticketing Area	Security Checkpoint	Boarding Lounge	Baggage Claim Hall	Arrival Hall	VIP Lounge
Combination Ⅰ	First floor	√	√	√	√	√	√	√
	Second floor				√			√
Combination Ⅱ	First floor	√	√			√	√	√
	Second floor			√	√			√
Combination Ⅲ	First floor					√	√	√
	Second floor	√	√	√	√			√

4.1.1 Combination Ⅰ : Boarding Lounge Located in the Upper Part

In this combination, the departure hall, check-in area, security checkpoint, arrival hall, baggage claim area, VIP lounge, and remote-stand boarding lounge are all located on the first floor. The nearby-stand boarding lounge is on the second floor, connected via a corridor bridge. The departure and arrival processes are closely integrated. This arrangement is mature and widely used domestically.

In the first-floor space of this combination, the arrival hall, departure hall, check-in waiting area, and security waiting area are arranged side by side. The long and tall front hall creates a grand and varied space with a rich and versatile layout.

On the second floor, only the nearby-stand boarding lounge is set up, leaving more space which provides more possibilities for the integration of characteristic spaces. For example, it can incorporate elements such as a light court, exhibition hall, leisure corridor, or ecological garden to enrich the spatial changes and increase the interests of small airports and terminal spaces. The terminal building shapes in Combination Ⅰ are diverse, commonly taking forms like rectangles, triangles, and circles. (Figs. 4-1, 4-2)

Cases of Combination Ⅰ

In Combination Ⅰ, the boarding lounge is located in the upper part, allowing for the addition of innovative spaces in the double-height hall, such as lighting patios, ecological courtyards, or overlapping water landscapes. These additions increase spatial diversity and promote visual communication among travelers.

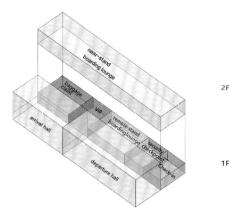

Figure 4-1 Axonometric drawing of Combination Ⅰ

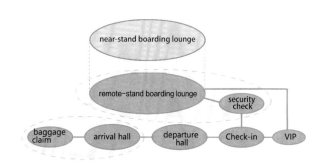

Figure 4-2 Spatial relationship diagram of Combination Ⅰ

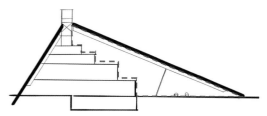
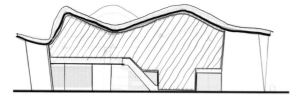

Figure 4-3 Sectional diagram of Datong Yungang Airport   Figure 4-4 Sectional diagram of Zaragoza New Airport

      For example, a lighting patio is added to the double-height hall at Shangrao Sanqingshan Airport in Jiangxi, reflecting the concept of "new rain on the empty mountain and ripples in the water," making the overall space bright, atmospheric, transparent, and fresh. Another excellent domestic example is Datong Yungang Airport in Shanxi, China, which integrates land-side lanes under a triangular frame, and functional spaces on each level of the terminal building (Fig. 4-3). It also incorporates numerous balcony spaces, enhancing sight communication and spatial mobility between different areas. The new airport in Zaragoza, Spain, features a large, wavy interleaved roof structure, with the boarding lounge set in the upper part and other functional spaces on the first floor. The variety of roof forms creates rich changes in the space on the waiting floor and the lobby on the first floor. (Fig. 4-4)

4.1.2 Combination II : Security Check and Boarding Lounge Located in the Upper Part

      The arrival hall, baggage claim hall, VIP lounge, remote-stand boarding lounge, and check-in hall are all situated on the first floor. The security checkpoint and near-stand boarding lounge are on the second floor, connected via a corridor bridge. The departure and arrival processes are interconnected. This combination is less used in China and more common abroad, with diverse terminal shapes.

      The characteristics of Combination II are: ① The departure and arrival flows are loosely linked, with only the check-in area on the first floor. Passengers who check in online can directly depart from the second-floor departure hall. The passenger flow paths are clear and efficient. ② The departure and arrival halls are spatially independent, allowing for separate designs. ③ The departure companion flow can reach the second floor, enriching the spatial experience and visual communication. ④ The terminal is usually rectangular and long, rarely square. The first floor is expansive, and the second floor has diverse space in this combination, with abundant flow paths. ⑤ Due to the different floor locations of the departure and arrival halls, this combination requires considerable terminal height and floor depth. It is necessary to

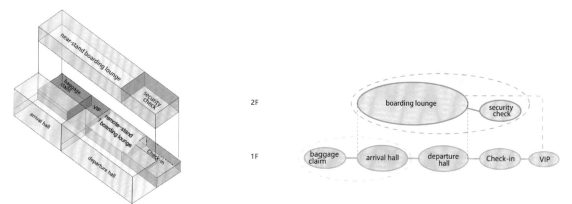

Figure 4-5 Axonometric diagram of Combination II   Figure 4-6 Spatial relationship diagram of Combination II

consider the design according to the depth and height limitations of the site. (Figs. 4-5, 4-6)

Cases of Combination II

The functional space on the first floor of the terminal building primarily serves arrivals, while the security check and boarding lounges are positioned on the second floor, dedicated mainly to departures with distinct and focused functionalities. The departure and arrival hall spaces intertwine vertically, integrating traditional functional spaces into the hall areas. Innovative spaces, such as leisure retail, double-height lighting atriums, exhibition galleries, and ecological courtyards, can also be incorporated into the flows, enriching the spatial experience and enhancing the airport's appeal, especially in smaller airport designs that emphasize local identity.

Beihai Fucheng Airport in Guangxi, China adopts the spatial layout of Combination II, placing the security check and near-stand boarding lounge on the second floor. The double-height spaces of the arrival and the departure halls are seamlessly integrated with functional flows, merging the halls with the baggage claim and security checkpoints. The circular arc shape of the halls with through-height spaces on both floors enhances spatial transparency and fluidity. (Fig. 4-7)

Figure 4-7 Sectional diagram of Beihai Fucheng Airport

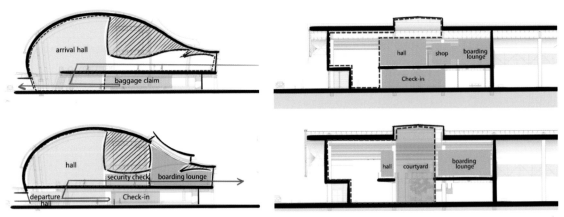

Figure 4-8 Sectional composition diagram of San José Mineta International Airport

Figure 4-9 Sectional composition diagram of La Araucanía Airport

San José Mineta International Airport in the United States aligns the departure and arrival halls side by side with double-height spaces, encouraging spatial interaction. The terminal's roof adopts an overall design treatment with rounded arc shapes, enhancing the dynamic and engaging nature of the boarding lounges. (Fig. 4-8)

At La Araucanía Airport in Chile, the arrival hall on the first and second floors are interconnected, while the departure and security checkpoints are vertically separated. Courtyard spaces are integrated within the process flow, creating a bright and open terminal atmosphere. Glass courtyards divide the main spaces, incorporating a sense of freshness and natural light into the halls and boarding lounges, ensuring a pleasant and visually stimulating environment. These spatial variations create layered experiences, diversifying terminal layouts and offering passengers enhanced journey experiences. (Fig. 4-9)

4.1.3 Combination III : Same-Level Entry and Exit, Layered Departure and Arrival

Combination III places the arrival hall and baggage claim hall on the first floor, while the second floor accommodates the departure hall, check-in area, security checkpoint, and boarding lounge connected by corridor bridges. This combination enhances the clarity and independence of both arrival and departure processes, with minimal functional space on the first floor and a larger footprint on the second floor. The terminal shapes vary widely.

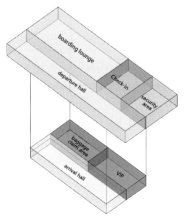 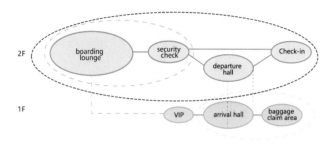

Figure 4-10 Axonometric diagram of Combination III

Figure 4-11 Spatial relationship diagram of Combination III

The characteristics of Combination III are: ① The functional spaces for departure and arrival processes are completely independent; ② The first floor has fewer functions with minimal spatial requirements, allowing for potential mezzanine spaces based on site conditions; ③ The second floor accommodates more diverse functional areas, requiring greater spatial depth; ④ Terminal shapes vary, mainly featuring squares, rectangles, or circular forms, with high demands for second-floor spatial height and depth. (Figs. 4-10, 4-11)

Cases of Combination III

The functional space on the first floor is mainly dedicated to the arrival process, while the second floor focuses on departures, ensuring clear process divisions within the spatial layout.

Liuzhou Bailian Airport in Guangxi, China adopts the standard layout of Combination III, incorporating a double-height courtyard within the departure process. This addition not only enhances visual and spatial connectivity with the arrival level, but also enriches the experience for departing passengers. Structurally, the departure and arrival halls are positioned along the short sides of the terminal with significant depth, while the departure waiting area extends along the terminal's long side to maximize space usage. (Fig. 4-12)

Kaunas Airport in Lithuania features double-height design for both arrival and departure halls. Elevated second-floor spaces host commercial areas positioned above, creating mezzanine commercial spaces with views of both departure and arrival halls. This commercial setup maintains close connectivity with the landside flow while allowing visual engagement with the airside, enriching the overall spatial experience in the terminal and enhancing interaction between passengers and the terminal during

Figure 4-12 Sectional composition of Liuzhou Bailian Airport

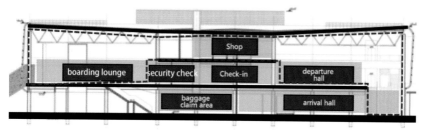

Figure 4-13 Sectional composition of Kaunas Airport

departure processes. The terminal's profile is dynamically shaped, integrating functions seamlessly. (Fig. 4-13)

4.1.4 Advantages, Limitations, and Scenarios for Three Combination Types

The three combination types have unique advantages and corresponding limitations. Different combination methods have different application scenarios.

Table 4-2 Advantages, Limitations, and Scenarios for Three Types of Combinations

Composition Type	Combination I : Boarding Lounge Located in the Upper Part	Combination II : Security Check and Boarding Lounge Located in the Upper Part	Combination III : Same-Level Entry and Exit, Layered Departure and Arrival
Advantages	1. The first-floor lobby is spacious and grand 2. Ample blank space on the second floor can be used to create special spaces 3. Diverse forms: rectangular, circular, triangular	1. The first and second floors are evenly distributed 2. Varied space and diverse flow lines 3. Diverse forms: rectangular, circular	1. The first-floor area is small with abundant blank spaces 2. The second floor offers diverse spaces and rich flow lines 3. Diverse forms: rectangular (square), circular
Limitations	1. The first floor's width is large, depth is moderate 2. Requires a large site area	1. Large depth requirement for the site 2. High terminal building height requirement	1. Requires a large site area 2. Terminal building requires large width and depth 3. High terminal building height requirement
Scenarios	1. Ample site area with sufficient width or depth 2. Suitable for rectangular, circular, or triangular terminal forms 3. Low terminal building height requirement	1. Moderate site area 2. Suitable for rectangular or circular terminal forms 3. Higher terminal building height	1. Ample site area 2. Suitable for rectangular (square) or circular terminal forms 3. Higher terminal building height

4.2 Spatial Scale Study of One-and-a-Half-Story Terminal

Spatial scale is one of the key factors influencing passenger experience. The design of spatial proportions and scales not only affects passengers' psychological perceptions, but also influences their visual impressions. From the perspective of psychological impact, larger spaces tend to evoke feelings of awe, emptiness, and alienation, while smaller spaces can create a sense of comfort, intimacy, or even oppression.

Table 4-3 Visual Perception of Space with Different *D/H* Values

Spatial Scale Diagram	H / D	H / D	H / D
D/H	D/H = 1	D/H = 2	D/H = 3
Elevation Angle	45°	27°	18°
Visual Perception	Resulting in difficult visual observation, and spatial details are not sufficiently presented	Providing a comfortable viewing angle, allowing passengers to obtain information within the space	Providing a comfortable viewing angle, allowing passengers to gather information from various elements arranged within such a large space and tell directions

4.2.1 Spatial Scale Study of Departure and Arrival Halls

4.2.1.1 Plane Scale

The departure and arrival halls are the most important public spaces within an airport terminal, where a large volume of arriving and departing passengers gather and move. After researching the plane scale of a large number of cases, the authors found that the departure hall of regional airport is generally similar to the width of the terminal building, with a depth of 15 to 24 meters. The ratio of the depth to the width is mostly concentrated in 1 : 2 to 1 : 4.

Table 4-4 Plane Scale of Entrance Hall of Several Regional Airports

Airport Name	Entrance Hall Area (m^2)	Width (m)	Depth (m)
Shennongjia Airport, China	772	45	15
Datong Yungang International Airport, China	937	54	16
Xizang Dingri Airport, China	790	50	15
Cangyuan Washan Airport, China	1,500	80	18

continued

Airport Name	Entrance Hall Area (m²)	Width (m)	Depth (m)
Lancang Jingmai Airport, China	910	64	13
Bengbu Airport, China	2,088	104	24
Shangrao Sanqingshan Airport, China	1,750	92	18
Yueyang Sanhe Airport, China	700	40	18
Hulunbuir Hailar Airport, China	2,880	116	24
Nelson Airport, New Zealand	530	42.4	12.5
The new airport of Georgia	430	28	15.2
La Araucanía Airport, Chile	570	60	12

4.2.1.2 Sectional Scale

According to the theory proposed by Peng Yigang and Ashihara Yoshinobu, different spatial forms bring different psychological and visual feelings to people. In sectional views, different D/H values provide varying spatial experiences. This book categorizes terminal building sections based on these values.

Table 4-5 D/H Value of Departure/Arrival Hall of Several Regional Airports (in meters)

Airport Name	Check-in Counter Layout	Depth (D)	Maximum Indoor Height (H_1)	Minimum Indoor Height (H_2)	Average Indoor Height (H_3)	D/H_3
Nelson Airport, New Zealand	Front Arrangement	11.6	8	6	7	1.65
San José Mineta International Airport (departure hall), USA	Front Arrangement	14.4	12.8	8.1	10.45	1.37
San José Mineta Internatioanl Airport (arrival hall), USA	Front Arrangement	14.4	3.9	3.9	3.9	3.7

continued

Airport Name	Check-in Counter Layout	Depth (D)	Maximum Indoor Height (H_1)	Minimum Indoor Height (H_2)	Average Indoor Height (H_3)	D/H_3
Kaunas Airport (departure hall), Lithuania	Front Arrangement	21	6.8	6.8	6.8	3.1
Kaunas Airport (arrival hall), Lithuania	Front Arrangement	14	3.3	3.3	3.3	4.2
La Araucanía Airport, Chile	Front Arrangement	12	8	7.2	7.6	1.6
Yueyang Sanhe Airport, China	Front Arrangement	18	10.4	8.3	9.35	1.9
Shangrao Sanqingshan Airport, China	Front Arrangement	18	11.9	10.4	11.15	1.6
Hulunbuir Hailar Airport, China	Front Arrangement	21	11.8	6.3	9.05	1.8
Xizang Dingri Airport, China	Front Arrangement	13	9.4	8.6	9	1.4
Bengbu Airport, China	Front Arrangement	29	21.5	12.5	17	1.7
Lancang Jingmai Airport, China	Front Arrangement	16.9	16.9	11.6	14.25	1.2
Cangyuan Washan Airport, China	Front Arrangement	18.8	13.5	11.8	12.65	1.5

1. Combinations I and II: Parallel Style

In both Combinations I and II, the departure and arrival halls of the terminal building share a large space with a net height right below the roof ceiling, so their sectional shapes are basically the same. (Fig. 4-14)

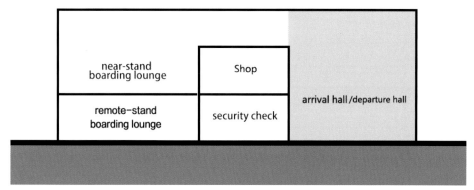

Figure 4-14 Combinations I and II : sectional composition of parallel departure/arrival halls

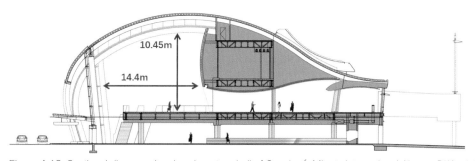

Figure 4-15 Sectional diagram showing departure hall of San José Mineta International Airport, D/H = 1.37

1) Parallel Departure and Arrival Halls (D/H < 1.5)

When the D/H value is less than 1.5, the hall's sectional depth and net height are too close, creating a sense of enclosure. This may make travelers feel that their vision is not wide enough, which can impact their ability to obtain information and tell location. (Fig. 4-15)

2) Parallel Departure and Arrival Halls (1.5 < D/H < 2.5)

When the D/H value is between 1.5 and 2.5, the sectional shape of the hall resembles a horizontal rectangle. In this type of space, the sense of enclosure is relatively reduced. As the D/H value increases, the space gradually creates a sense of broadness, allowing travelers to more easily view signs and obtain information. (Figs. 4-16 ~ 4-18)

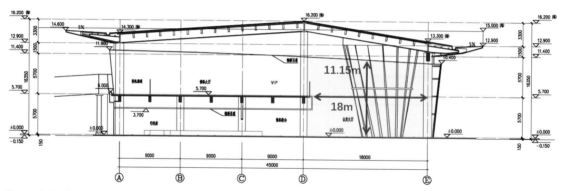

Figure 4-16 Sectional diagram showing departure/arrival hall of Shangrao Sanqingshan Airport, D/H = 1.6

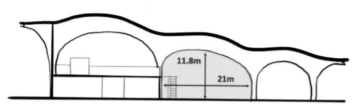

Figure 4-17 Sectional diagram showing departure/arrival hall of Hulunbuir Hailar Airport, D/H = 1.8

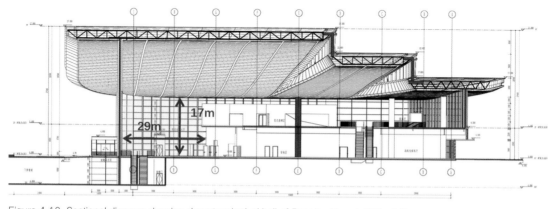

Figure 4-18 Sectional diagram showing departure/arrival hall of Bengbu Airport, D/H = 1.7

2. Combination III: Layered Style

1) Layered Style Departure Hall (2.5 < D/H < 3.5)

In Combination III, the arrival hall is generally located on the first floor to better connect with the baggage claim hall, while the departure hall is located on the second floor. Since the departure and arrival halls in this combination do not share the same interior net height, the D/H value is larger than that in Combinations I and II, typically between 2.5 and 3.5. Additionally, the net height of this type of space is usually around 6 to 9 meters, which is more conducive to passengers obtaining visual information and provides a more comfortable spatial experience. (Figs. 4-19, 4-20)

2) Layered Style Arrival Hall (3.5 < D/H < 5)

Layered style arrival halls are usually set on the first floor, with a net height from the finished ground surface to the ceiling below the upper floor, usually around 3.5 to 5 meters. This height is relatively smaller than the net height of the departure hall above, resulting in a D/H value between 3.5 and 5. While

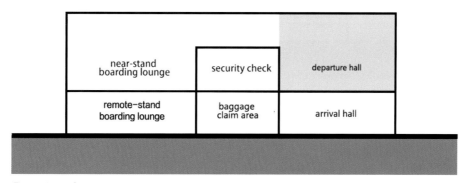

Figure 4-19 Combination III: sectional composition of layered departure hall

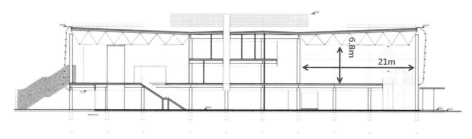

Figure 4-20 Sectional diagram showing departure hall of Kaunas Airport, D/H = 3.1

a smaller *D/H* value is advantageous for obtaining visual information, the limited net height may create a sense of oppression. If the net height is too low, it can negatively impact the visual experience when obtaining information. (Figs. 4-21, 4-22)

Based on the above analysis, the depth of departure and arrival halls is typically 15 to 24 meters. For parallel halls in Combinations I and II, it is recommended that the average indoor net height is between 10 to 16 meters. For layered halls in Combination III, it is recommended that the upper space have an average net height of 6 to 10 meters, and the lower space be at least 3.6 meters. At this scale, travelers can experience the openness of the space and conveniently view signs and information.

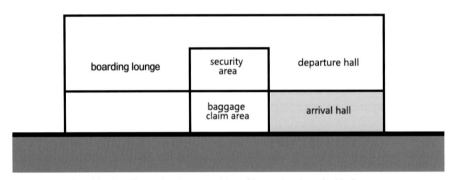

Figure 4-21 Combination III: sectional composition of layered style arrival hall

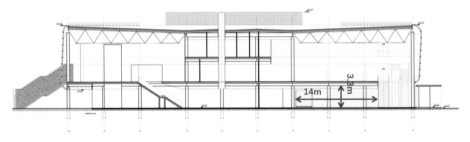

Figure 4-22 Sectional diagram showing arrival hall of Kaunas Airport, *D/H* = 4.2

4.2.2 Spatial Scale Study of Security Checkpoint

4.2.2.1 Plane Scale

The width of security checkpoint in a one-and-a-half-story terminal is generally determined by the number of screening machines (each machine has a width of 5 meters), and the depth is usually between 16 and 20 meters.

Table 4-6 Comparison of Number of Screening Machines, Width, and Depth of Security Checkpoint of Several Regional Airports

Airport Name	Number of Security Screening Machines	Width (m)	Depth (m)
Shennongjia Airport, China	2	10	16
Datong Yungang International Airport, China	4	18	14
Xizang Dingri Airport, China	4	18	18
Cangyuan Washan Airport, China	3	13.3	20
Lancang Jingmai Airport, China	5	20	18
Bengbu Airport, China	8	34	18
Shangrao Sanqingshan Airport, China	3	15	10
Yueyang Sanhe Airport, China	2	10	10
Hulunbuir Hailar Airport, China	8	16	21
The new airport of Georgia	3	16.5	10
La Araucanía Airport, Chile	2	9	10

4.2.2.2 Sectional Scale

The security checkpoint needs to accommodate a large number of travelers quickly, so it requires sufficient horizontal width and an adequate height to avoid a depressing feeling. Security checkpoints vary in height depending on their location, and this section discusses both lower-level and upper-level types.

1. Combination I: Lower-Level Security Checkpoint

In Combination I, the lower-level security checkpoint has commercial areas or near-stand boarding lounges on the second floor. To meet the basic spatial needs of passengers, the net height needs to be set at 3 to 4 meters. For instance, the security checkpoint at Shangrao Sanqingshan Airport has a net height of 3.7 meters, while at Bengbu Airport, it is 3.6 meters. Thus, the depth-to-height ratio ranges from 4 to 6. (Figs. 4-23 ~ 4-25)

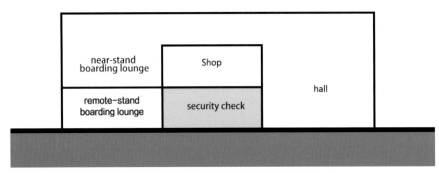

Figure 4-23 Sectional composition of lower-level security checkpoint

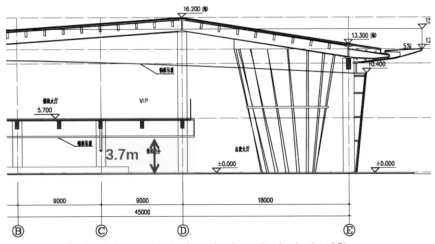

Figure 4-24 Sectional diagram showing lower-level security checkpoint of Shangrao Sanqingshan Airport, Net height (H) = 3.7 m

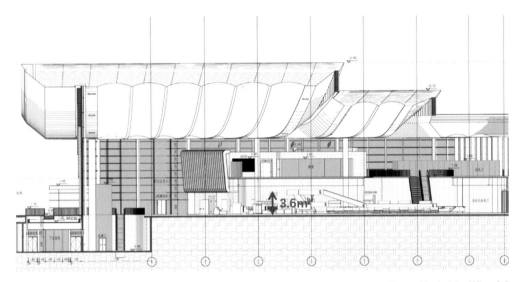

Figure 4-25 Sectional diagram showing lower-level security checkpoint of Bengbu Airport, Net height (H) = 3.6 m

2. Combinations II and III: Upper-Level Security Checkpoint

In Combinations II and III, the upper-level security checkpoint has no floor above it, so the net height can be calculated directly to the ceiling. The height of separation plates is usually around 2.5 meters, which meets the need for clear sign identification. Therefore, the height-to-depth ratio ranges from 6 to 8. For instance, the security checkpoints at Hulunbuir Hailar Airport and San José Mineta International Airport, located on the second level platforms, have net heights extending up to the roof, only requiring the separation plates to meet the signage recognition height. (Figs. 4-26, 4-27)

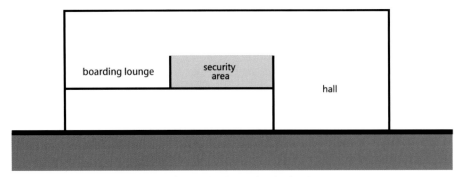

Figure 4-26 Sectional composition of upper-level security checkpoint

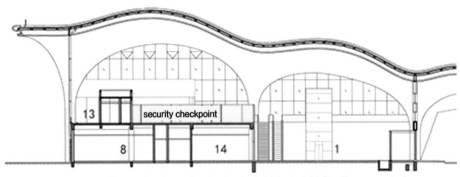

Figure 4-27 Location of upper-level security checkpoint in Hulunbuir Hailar Airport terminal

  Based on analyses above, it is recommended that lower-level security checkpoints in Combination I have an average net height of 3 to 4 meters to facilitate quick passenger traffic. For upper-level security checkpoints in Combinations II and III, the height of separation plates should be controlled between 2.5 and 3 meters to ensure clear visibility of signage.

4.2.3 Spatial Scale Study of Boarding Lounge

4.2.3.1 Plane Scale

The layout of the boarding lounge is related to the terminal's flow organization. The depth of the boarding lounge typically ranges from 12 to 20 meters, and the width generally matches the terminal's width, with a width-to-depth ratio between 1:1 and 1:3. (Table 4-7)

Table 4-7 Comparison of Waiting Mode and Boarding Lounge Plane Scale of Several Regional Airports

Airport Name	Boarding Lounge Area (m^2)	Waiting Mode	Width (m)	Depth (m)
Shennongjia Airport, China	550	Single side	45	12
Datong Yungang International Airport, China	1,400	Single side	110	12
Xizang Dingri Airport, China	430	Single side	20	18
Cangyuan Washan Airport, China	1,750	Single side	152	12
Lancang Jingmai Airport, China	1,900	Single side	180	11
Bengbu Airport, China	5,900	Single side	265	25
Shangrao Sanqingshan Airport, China	1,250	Single side	62	20
Yueyang Sanhe Airport, China	690	Single side	54	12.6
Hulunbuir Hailar Airport, China	2,020	Single side	117	18
Nelson Airport, New Zealand	720	Single side	28	26.2
The new airport of Georgia	620	Single side	32	18
La Araucanía Airport, Chile	1,050	Single side	95	14

4.2.3.2 Sectional Scale

The large space of the boarding lounge can be divided into seating and passage areas, both typically set within the same large space. The sectional design of the boarding lounge should not only ensure the comfortable spatial scale of the seating area, but also provide suitable viewing angles for passengers in the passage area.

In a one-and-a-half-story terminal, near-stand boarding lounges are usually located on the upper level, connecting with boarding bridges, while remote-stand boarding lounges are typically on the lower level, connecting with bus boarding. Since the spatial experience differs between the two levels, the two types will be discussed separately as follows:

1. Near-Stand Boarding Lounge

The visual and spatial feeling of the upper-level near-stand boarding lounge is more open because its net height is calculated from the finished surface of the second floor to the ceiling of the roof. Through a comparative study of data from several cases, it can be found that the D/H value is concentrated in the range of 2 to 3.5, and the net height is around 5 to 9 meters. This provides passengers in the passage area with a broad view to identify information and judge their position, while passengers in the seating area can feel a sense of enclosure and stability due to the certain spatial range limitation provided by the seats. (Table 4-8, Figs. 4-28 ~ 4-31)

Table 4-8 Comparison of Waiting Mode, Height, and D/H Value of Near-Stand Boarding Lounges of Several Regional Airports (in meters)

Airport Name	Waiting Mode	Depth (D)	Maximum Interior Height (H_1)	Minimum Interior Height (H_2)	Average Interior Height (H_3)	D/H_3
Nelson Airport, New Zealand	Single side	30.4	10.5	6.8	8.65	3.4
Kaunas Airport, Lithuania	Single side	19.6	6.8	6.8	6.8	2.9
La Araucanía Airport, Chile	Single side	10.4	5.6	4.8	5.2	2
Yueyang Sanhe Airport, China	Single side	16.4	5.6	4.8	5.2	3.1
Shangrao Sanqingshan Airport, China	Single side	21.2	6.2	5.4	5.8	3.65
Hulunbuir Hailar Airport, China	Single side	20.5	10.6	5.1	7.85	2.6

continued

Airport Name	Waiting Mode	Depth (D)	Maximum Interior Height (H_1)	Minimum Interior Height (H_2)	Average Interior Height (H_3)	D/H_3
Xizang Dingri Airport, China	Single side	20.6	16.1	11	13.55	1.8
Bengbu Airport, China	Single side	21.8	6.9	6.3	6.6	3.3

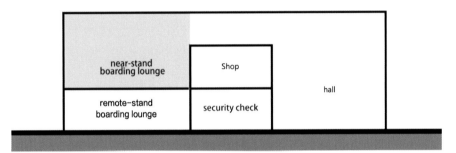

Figure 4-28 Sectional composition of upper-level near-stand boarding lounge

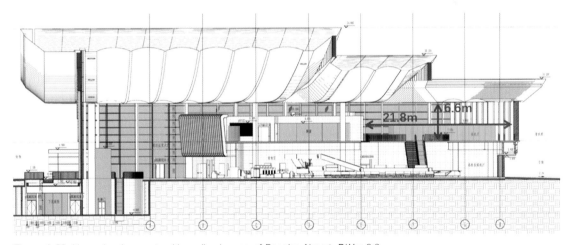

Figure 4-29 Upper-level near-stand boarding lounge of Bengbu Airport, D/H = 3.3

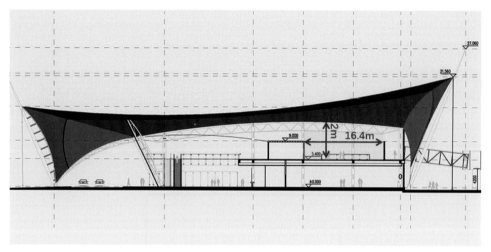

Figure 4-30 Upper-level near-stand boarding lounge of Yueyang Sanhe Airport, D/H = 3.1

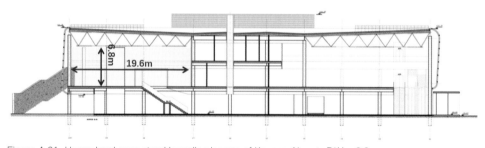

Figure 4-31 Upper-level near-stand boarding lounge of Kaunas Airport, D/H = 2.9

2. Remote-Stand Boarding Lounge

Due to the floor height limitation, the net height of the lower-level remote-stand boarding lounge is often between 3 to 5 meters. However, to ensure the spatial experience of passengers, its depth is usually not too large, while the area of seating and passage zones is smaller than that of the upper-level near-stand boarding lounge, so the overall D/H can still be maintained at about 2.5 to 4. (Table 4-9) (Figs. 4-32 ~ 4-34)

Table 4-9 Waiting Mode, Height, and D/H Value of
Remote-Stand Boarding Lounges at Several Regional Airports (in meters)

Airport Name	Waiting Mode	Depth (D)	Maximum Interior Height (H_1)	Minimum Interior Height (H_2)	Average Interior Height (H_3)	D/H_3
Yueyang Sanhe Airport, China	Single-side	18	4.2	4.2	4.2	4.2
Shangrao Sanqingshan Airport, China	Single-side	9	3.7	3.7	3.7	2.4
Bengbu Airport, China	Single-side	9.5	3.6	3.6	3.6	2.6

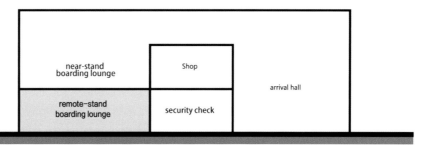

Figure 4-32 Sectional composition of lower-level remote-stand boarding lounge

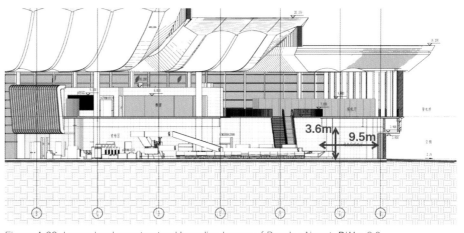

Figure 4-33 Lower-level remote-stand boarding lounge of Bengbu Airport, D/H = 2.6

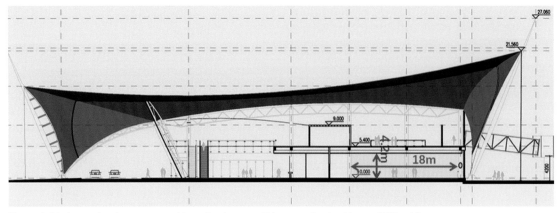

Figure 4-34 Lower-level remote-stand boarding lounge of Yueyang Sanhe Airport, *D*/*H* = 4.2

According to the comprehensive analyses above, the common depth of the boarding lounge is generally 12 to 20 meters. Therefore, it is recommended that the average net height of the upper-level near-stand boarding lounge be 5 to 9 meters, and that of the lower-level remote-stand boarding lounge be 3 to 5 meters, to ensure the openness of travelers' visual fields and facilitate the identification of information and of location.

4.2.4 Spatial Scale Study of Baggage Claim Hall

4.2.4.1 Plane Scale

The plane scale of the baggage claim hall of the terminal building is related to the type of baggage belt, arrangement and column grid. The depth of the baggage claim hall is generally about 15 to 30 meters, and the *D*/*W* value is concentrated in 1 : 2 to 1 : 4. (Table 4-10)

Table 4-10 Area of Baggage Claim Hall and Number of Baggage Carousels in Several Regional Airports

Airport Name	Area of Baggage Claim Hall (m²)	Number of Carousels	Width (m)	Depth (m)
Shennongjia Airport, China	270	1	17	22
Datong Yungang Airport, China	620	2	40	14
Xizang Dingri Airport, China	260	1	22	16
Cangyuan Washan Airport, China	760	2	31	18
Lancang Jingmai Airport, China	1,050	2	42	24
Bengbu Airport, China	1,500	2	51	24
Shangrao Sanqingshan Airport, China	860	2	38.5	21.5
Yueyang Sanhe Airport, China	615	2	30	20
Hulunbuir Hailar Airport, China	1,280	3	60	20
Nelson Airport, New Zealand	460	1	15.2	30.6
The new airport of Georgia	540	2	18.6	27.8
La Araucanía Airport, Chile	1,100	2	60	22

4.2.4.2 Sectional Scale

In the baggage claim hall, the primary purpose of travelers is to obtain information quickly and collect their baggage on the carousel as soon as possible, so the D/H value of baggage claim hall is suggested more than 3. After analyzing a large number of cases of one-and-a-half-story terminals, two types will be discussed as follows.

Table 4-11 *D/H* Value of Baggage Claim Halls of Different Regional Airports (in meters)

Airport Name	Depth (*D*)	Maximum Interior Height (*H₁*)	Minimum Interior Height (*H₂*)	Average Interior Height (*H₃*)	*D/H₃*
San José Mineta International Airport, USA	30	5.1	5.1	5.1	5.9
Kaunas Airport, Lithuania	13.2	3.3	3.3	3.3	4
La Araucanía Airport, Chile	28	5	5	5	5.6
Nelson Airport, New Zealand	16.8	4.2	4.2	4.2	4
Yueyang Sanhe Airport, China	18	4.2	4.2	4.2	4.3
Hulunbuir Hailar Airport, China	16.8	5.2	5.2	5.2	3.2
Xizang Dingri Airport, China	17.4	2.6	2.6	2.6	6.7
Bengbu Airport, China	20.7	3.6	3.6	3.6	5.8
Lancang Jingmai Airport, China	24.3	4.5	4.5	4.5	5.3
Cangyuan Washan Airport, China	18	4.8	4.8	4.8	3.8

When the *D/H* value is greater than 3.5, passengers in the baggage claim hall find it easier to distinguish flight and baggage information. The baggage carousel area also provides a spacious boarding lounge to accommodate passengers queueing for luggage collection. In such halls, the *D/H* value is often between 3.5 and 8. However, given the limited ceiling height of the lower level, excessive depth can make people feel a sense of oppression and discomfort. Thus, it is essential to avoid *D/H* values too large in this kind of space design. (Figs. 4-35 ~ 4-37)

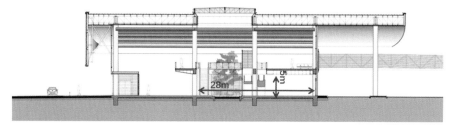

Figure 4-35　Sectional diagram showing baggage claim hall of La Araucanía Airport, D/H = 5.6

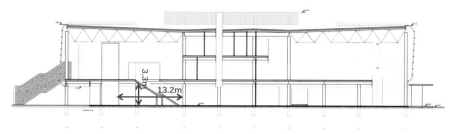

Figure 4-36　Sectional diagram showing baggage claim hall of Kaunas Airport, D/H = 4

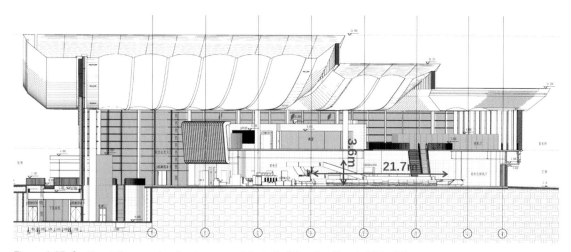

Figure 4-37　Sectional diagram showing baggage claim hall of Bengbu Airport, D/H = 5.8

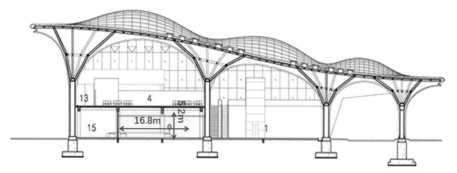

Figure 4-38 Sectional diagram showing baggage claim hall of Hulunbuir Hailar Airport, *D/H* = 3.2

    When *D/H* value is less than 3, the depth of the baggage claim hall will be shortened, which affects passengers' access to baggage information to a certain extent. If *D/H* value is less than 2, the net height and depth of the space are close to each other, which not only leads to a restricted visual field and a sense of spatial depression, but also makes the waiting space of baggage carousel more constricted.

    According to the comprehensive analyses above, the common depth of the baggage claim hall is 15 to 30 meters. Therefore, it is recommended that the average net height be between 3.6 and 5 meters, not only to facilitate access to baggage information for passengers, but also to ensure the space comfort of the carousel waiting area.

INNOVATION

DESIGN INNOVATION OF REGIONAL AIRPORT TERMINAL BUILDING

With the development of society and the economy, the demand for the functional quality and spatial experience of terminal buildings continues to rise. China's strong support for the construction of civil aviation hubs provides new opportunities and challenges for the construction of small- and medium-sized terminals. In this context, improving the overall creative level and exploring the personalized expression of regional culture has gradually become an important research direction in terminal design. Innovative research on terminal building design can promote innovation of terminal building forms and provide new ideas and methods for design theory, function, and technology of terminal buildings. This helps meet modern needs and create terminal buildings with characteristics of the era, region, culture, and good experience. Therefore, innovative design for terminal buildings has become a significant task in the current era.

5.1 Problems in Design of Terminal Building

Although one-and-a-half-story terminals have a promising development prospect, the author has found during research and design practice that in recent years, the design and construction of China's regional airport terminals have exposed many problems while experiencing rapid quantitative growth. Issues such as overlooking the temperament of transportation architecture, lack of aesthetics, superficial design expressions, and severe homogenization have been observed. These problems and typical cases are analyzed below.

① **Ignoring Terminal Building's Public Attributes and Temperament as Transportation Architecture**

Among existing small regional airports, some early projects lack the experience necessary for transportation building design. This results in terminal buildings that lack the characteristics of public buildings and transportation architecture in terms of facade, volume, materials, and details, presenting the issue of "unlike airports."

② **Bizarre Shapes and Lack of Aesthetics**

The second type of problem appeared mainly in the early stages of the construction boom of small- and medium-sized regional airports after 2000. This issue is primarily reflected in terminal designs that seek novelty and uniqueness, often resulting in imbalanced shapes and volumes, rough techniques, and disproportionate scales. The final terminal designs often appear bizarre and lack a consensus aesthetic.

③ **Abuse of Collage and Stacking for the Sake of Region**

The third type of issue is mostly seen during the period when airport and terminal design began emphasizing the integration of regional characteristics and culture. However, these efforts are often superficial, manifested through excessive use of regional symbols, cultural elements, and colors. This approach, termed "for the sake of region," lacks in-depth excavation and extraction of the local cultural context, making it difficult to achieve a high level of unity between culture and form, space, and experience.

④ Serious Homogenization and Lack of Diversity

The fourth type of issue has become more prominent as small- and medium-sized airports increasingly learn from the design experience of large airports. They often ignore regional culture and characteristics, blindly apply the design methods of large airport terminals and pursuing large roofs and streamlined forms. This leads to a "one size fits all" phenomenon, where airport terminal buildings across different regions, climates, and cultural atmospheres appear homogenized. This homogenization makes it difficult for regional airports to serve effectively as city business cards and tourism windows.

In summary, the design of small- and medium-sized terminals in China largely remains at a superficial level of visual expression, lacking essential innovation. Extreme regionalism and rampant internationalism in design styles have led to chaotic and disorderly terminal designs, resulting in bizarre and exaggerated terminal buildings. These buildings fail to serve as important "air portals," or to highlight the city's image and cultural characteristics. Addressing these issues urgently requires exploring positive and effective solutions, and the innovative design research for China's regional airport terminals is imperative.

5.2 Innovative Methods and Excellent Cases of Terminal Building Design

Compared to other countries, the development of civil aviation airports in China has been relatively slow. While international terminal design has advanced through numerous attempts and explorations, terminal design in foreign countries has been continuously developing, leading to breakthroughs and innovations in both design concepts and construction techniques, China's civil aviation hub construction started later and lacks comprehensive theoretical research on terminal design. Particularly, there is a scarcity of specialized studies from an architectural creation perspective that focuses on regional characteristics, and there is no established system of innovative design methods to guide the design of small- and medium-sized terminals. Therefore, this section examines successful international terminal design cases to summarize innovative design ideas and methods, providing references for the design practice of Chinese small- and medium-sized terminal buildings.

This section contains 4 Chinese airport cases and 27 cases in other countries and regions. By systematically summarizing these cases and drawing on their rich design experiences, we discuss the architectural form, spatial layout, and detailed construction of terminals in multiple dimensions. This analysis aims to provide inspiration and guidance for the design practice of small- and medium-sized terminals in China. The following section will focus on three aspects: external form, internal experience, and detailed construction, exploring the specific application of innovative design methods in terminal design.

5.2.1 Innovative Design of External Form

The external form of a terminal building is crucial in shaping its overall image and temperament. It reflects the terminal's design concept and is key to showcasing its appearance and highlighting regional cultural characteristics. With the continuous evolution of terminal design in the new era, increasing attention is paid to distinctiveness and originality, and a deep understanding of regional culture has become an important entry point for innovative terminal design.

Innovative concepts directly and effectively influence the design expression of a terminal's external form. These concepts are not only essential for showcasing regional cultural characteristics, but also represent a vital dimension of the design process. The innovative design of external form should go beyond mere formal exploration to delve into the underlying logic and creative methods. This section analyzes the design concepts of exemplary terminal practices, exploring the transition "from concept to form" and extracting innovative ideas applicable to external form design. The subsequent discussion will detail two innovative approaches: "starting from natural elements" and "starting from humanistic attributes."

5.2.1.1 Innovative Ideas Based on Natural Elements

The objective natural conditions of the terminal site, including geographical location, topography, climate, and other factors, often serve as starting points for small- and medium-sized terminal design. By rethinking the relationship between terminal buildings and the natural environment, designers can create innovative forms that are suitable for the local climate and integrated into the natural environment with regional characteristics.

Terminal building design ideas inspired by natural elements are diverse and creative. They include bionic buildings inspired by natural forms, landscape architecture that harmonizes with natural terrain, and new building forms adaptive to local climates. Different creative approaches ultimately present a rich variety of terminal building externals. This section summarizes three design techniques for innovative external form: drawing on topography, echoing natural climates, and adopting bionic abstract forms.

1. Drawing on Topography

Terminal design should respect the original topography, emphasizing the integration of the natural environment with the architectural form. Unique natural topography often becomes a key starting point of the design concept. By abstracting and refining local geomorphic features, designers can integrate these elements into the terminal's shape, materials, or textures, giving the building a distinctive form.

Case Study: Lleida Airport in Spain
The building texture echoes the wheat field

Lleida Airport is a small international airport located in northeastern Spain. Its terminal building is a single-story structure with a floor area of about 5,000 square meters, surrounded by the vast natural scenery of the Lleida highlands.

Due to the location and small size of the project, the architects aimed to balance creating an eye-catching form with maintaining iconic features. They adopted an integrated design approach, merging the tower with the main terminal through two strong curves to create a cohesive visual effect.

The roofs of the arrival hall and ancillary facilities extend upward to form the outer skin of the tower, unifying horizontal and vertical elements under two large "green carpets." The airport

Figure 5-1 Photo of Lleida Airport

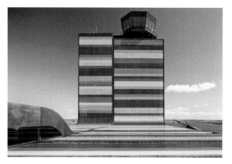

Figure 5-2 Material and texture of the exterior facades of Lleida Airport

Figure 5-3 The terminal building of Lleida Airport integrates with surrounding environment

roof employs various materials, such as vegetation, wood, and metal panels, creating textures and tones that harmonize with the surrounding landscape, seamlessly blending the airport with the surrounding natural environment.

The Lleida Airport terminal, with its simple, expansive, and recognizable form and the variations in external texture reflecting the local landscape, presents an image that is both understated and locally iconic. The design considers functionality and aesthetics while fully embracing the harmonious coexistence of the airport and its surrounding environment. (Figs. 5-1 ~ 5-3)

Case Study: Carrasco International Airport in Uruguay
The undulating design blends with the hilly surroundings

Carrasco International Airport, located in Montevideo, the capital of Uruguay, is the only airport in the country that operates international flights year-round, with an annual passenger throughput reaching 2 million person-times. The new terminal covers an area of 45,000 square meters and consists of two floors, with the arrival hall on the first floor and the departure hall on the second floor.

The airport is situated in a dune area along the coast of Montevideo, and the gentle curves and understated nature of the dunes greatly inspired the designers. The airport features an indoor-outdoor integrated design, with the roof conceived as a complete lightweight shell that spans 366 meters across the terminal, connecting directly to the ground at both ends. The architectural form not only echoes the surrounding dune landscape in its external appearance, but also closely

Figure 5-4 Bird eye view of Carrasco International Airport

Figure 5-5 Exterior facade of Carrasco International Airport

aligns with the texture of the dunes in its material choices and detailing.

The roof structure is composed of three materials: a white thermoplastic membrane for the roof, external panels, and a white titanium-based vinyl membrane for the ceiling. By carefully controlling the color and detailing of these materials, the design achieves a seamless, minimalist shell that integrates harmoniously with natural environment. (Figs. 5-4, 5-5)

The design pays close attention to the local tradition of saying goodbye to family and friends before departure, so special emphasis is placed on the shaping of public areas, including the central hall, departure hall, and outdoor terrace. Through the rational arrangement of functions and layout, the design integrates the interior spaces organically, bringing travelers a rich spatial experience beneath the simple exterior shape. The whole building is rooted in its surroundings and is given a contemporary feel through its treatment of space, function and structure.

2. Adapting to Natural Climate

Various regional climatic characteristics play a significant role in shaping architectural forms. Different regions have unique climatic conditions, so considering and responding to these conditions often inspire terminal design. Additionally, creating a pleasant indoor environment and reducing building energy consumption are core goals of architectural innovation. With the increasing standards for green and energy-efficient design, it has become particularly important for terminal buildings to adapt to local climatic characteristics.

Case Study: Banyuwangi International Airport in Indonesia
Wide overhanging eaves provide shade and open design enhances ventilation

Banyuwangi International Airport is located in a small town in East Java, Indonesia, bordered by mountains to the west and the sea to the east. The airport covers a total area of about 20,000 square meters. The building is divided into two volumes, housing the departure and arrival halls separately to meet different functional needs.

The airport is surrounded by local rice paddies, and in the summer, the green fields are the first sight that greets travelers as they land. The terminal's roof is covered with green vegetation, blending seamlessly with the surrounding landscape when viewed from above, creating a

Figure 5-6 Perspective view of Banyuwangi International Airport

Figure 5-7 Sectional drawing of Banyuwangi International Airport terminal

harmonious coexistence with nature. (Fig. 5-6)

The wide eaves and the sloping roof of the airport terminal are inspired by local architectural features. The extensive eaves, extending over 4 meters, can block strong direct sunlight, effectively reducing indoor temperatures in summer. The sloping roof, combined with a green roof design, also absorbs heat. This architectural form not only reflects strong regional characteristic, but also incorporates traditional methods effective in coping with the tropical climate.

The facade adopts an open design, with vertical louvres enclosing the exterior to enhance air circulation within the terminal and control lighting. This approach also effectively introduces outdoor views, allowing visitors to get closer to nature while resting and waiting. The design utilizes atriums and side halls, combined with water features, to help create a cool indoor environment against the heat of the tropics. This design, which emphasizes openness and natural elements, gives the terminal a unique visual appeal. (Fig. 5-7)

Case Study: Guelmim Airport in Morocco
Simple volumes reduce building energy consumption,
and light atrium regulates the natural climate

Located in the southern part of the Little Atlas Mountains, near the northwestern edge of the Sahara Desert, Guelmim Airport serves as an important gateway to the desert. The airport integrates a nearby military airfield, and includes a terminal area of approximately 9,000 square meters. The project emphasizes simplicity, energy efficiency, flexibility, and scalability to cope with the extreme natural environment. The building uses simple rectangular volumes, harmoniously blending with the desert landscape, while the contrast between the straight lines of building and the curves of the sand dunes highlights the natural beauty of the desert. (Fig. 5-8)

To achieve good lighting performance and indoor temperature control in the desert climate, the terminal design adopts a two-layer shell structure. The inner layer is a glass curtain wall, while the outer layer consists of colorful perforated metal panels. These metal panels filter solar

Figure 5-8 Perspective view of Guelmim Airport terminal and its surroundings

Figure 5-9 Atrium curtain wall with perforated metal panels of Guelmim Airport terminal

radiation and create a buffer zone to reduce indoor temperature. The design of the metal panels draw inspiration from traditional Moroccan decorations and reflect the colors of the surrounding environment. (Fig. 5-9) Additionally, the atrium space inside the building enhances natural lighting and helps regulate the indoor microclimate, reducing energy consumption. The atrium's membrane structure balances shading and lighting needs. Overall, the design of Guelmim Airport fully considers the characteristics of the local climate, respects the natural environment, and pursues sustainable development, achieving a unity of functionality and aesthetics.

3. Adopting Bionic Abstract Form

Bionic design draws inspiration from the functional organization and morphological composition principles found in nature, researching the scientific construction rules of organisms and applying these principles to architectural design to achieve more efficient structural design and functional layout. In contemporary terminal design, bionic techniques have been widely adopted, becoming key methods for enriching architectural forms and enhancing design innovation.

Airports, closely associated with flight, movement, and speed, often find inspiration from the dynamic characteristics of organisms in nature. Birds, in particular, act as a primary example. The invention of airplanes owes much to study of bird flight, and bird postures are frequently reflected in the design of airport terminals.

Case Study: Gelendzhik Airport in Russia
Simulating a flying bird's turn

Gelendzhik Airport, designed by the Fuksas office, covers a total area of approximately 7,800 square meters, with an expected annual passenger capacity of over one million person-times.

Inspired by the turning movement of birds in flight, this design embodies the application of bionics in architecture. The building's form mimics the elegant turning posture of birds, creating a light and dynamic shape through smooth curves and dynamic structures. Due to its unique and innovative design, Gelendzhik Airport has become a natural landmark of the city, symbolizing the modern architectural concept of harmonious coexistence between humans and nature. (Fig. 5-10) The interior space design integrates local natural elements, using "sea and wind" as spatial themes and employing parametric design to create a poetic space. The roof, made of white triangular composite panels, simulates the dynamic scene of wind blowing over the sea and waves rolling, forming a visually appealing and poetic ceiling landscape. (Fig. 5-11)

Unlike traditional airport design models, Gelendzhik Airport's design utilizes innovative techniques to explore the balance between architecture and nature, function and aesthetics. This experimental design approach not only provides new ideas for future airport architectural design, but also offers fresh perspectives and possibilities for the entire construction industry.

Figure 5-10 Bird eye view of Gelendzhik Airport

Figure 5-11 Pattern of the ceiling at Gelendzhik Airport

Case Study: Lishui Airport in Zhejiang, China
Simulating a bird spreading its wings

Lishui Airport, designed by MAD Architects, is a one-and-a-half-story terminal with an area of approximately 12,000 square meters. Expected to be completed in 2024, it will serve over one million person-times annually.

The terminal design is inspired by the elegant posture of a bird spreading its wings. The design team constructed a structure that mimics bird wings, creating a light and fluid visual effect. The exterior stretches upward like a bird's wings, presenting a dynamic aesthetic ready for takeoff. (Figs. 5-12, 5-13) The integrated indoor-outdoor design harmonizes the form and function of the terminal. Its silver-white aluminum roof extends outward, creating a series of continuous eaves that provide shade and shelter from the rain in the drop-off area. A skylight at the peak of the roof effectively brings natural light into the interior, enhancing the brightness and comfort of the space. Respecting the principles of nature and bionics, the design meticulously creates a landmark building that combines artistic beauty and practicality.

Figure 5-12 Rendering of the facade of Lishui Airport terminal

Figure 5-13 Rendering of the lateral facade of Lishui Airport

Summary: This section elaborates on innovative approaches inspired by natural elements, extracting three design strategies: drawing on topography, adapting to natural climates, and adopting bionic abstract forms. These strategies introduce new perspectives and methods for terminal design, imbuing the designs with distinct regional characteristics and effectively addressing the prevalent issue of design homogenization. Furthermore, these strategies reinforce the harmonious relationship between terminals and the surrounding natural environment, enhancing the ecological adaptability of the buildings and making them exemplary models of environmentally friendly design.

5.2.1.2 Innovative Ideas Based on Humanistic Attributes

Humanistic elements play an indispensable role in airport terminal design, encompassing aspects such as era characteristics, social background, local historical context, and cultural customs. These elements collectively contribute to a design scheme that showcases distinct regional cultural characteristics, creating a space with a unique local atmosphere that serves as a window showing urban culture. Terminal designs that focus solely on basic functions while neglecting humanistic attributes often lack vitality. Conversely, designs that fully consider and integrate cultural elements can imbue the terminal with rich cultural connotations and spiritual value, evoking emotional resonance in passengers. Therefore, the in-depth exploration and precise expression of regional culture have become core issues in innovative terminal design.

This section, through case studies, refines the concentrated expression of cultural elements in the exterior design of airport terminals, highlighting two design techniques: "abstract extraction of cultural symbols," and "spatial metaphor of regional culture."

1. Abstract Extraction of Cultural Symbols

Cultural symbols, as carriers of traditional culture, typically manifest as visual images with spiritual and symbolic significance. In airport terminal design, these symbols are often reflected in traditional architectural styles, regional cultural totems, and the colors and forms associated with religious beliefs. In innovative design practices centered on humanistic attributes, designers commonly extract local cultural elements and abstract them into architectural forms or utilize unique space archetypes and materials from the local culture to create airport terminals with distinctive regional features.

Moreover, when using cultural symbols, designers should avoid simple replication and collage. Instead, they should abstract, deconstruct, and reinterpret these symbols. Such abstract expression can stimulate viewers' imagination and evoke passengers' empathy for the culture conveyed by the terminal building, representing an advanced and nuanced design approach.

Case Study: Terminal 2 of Datong Yungang International Airport in Shanxi, China
The roof reflects unique regional temperament of Shanxi

The Terminal 2 of Datong Yungang International Airport in Shanxi, China completed in 2013, has a total area of 10,854 square meters and an annual passenger throughput of up to 900,000 person-times, making it a representative small- and medium-sized regional airport in China. The design of the new terminal building introduces the "Datong Roof" concept, aiming to extract and reinterpret traditional Shanxi architectural elements and integrate them into modern airport terminal design through modernization and abstraction. This approach creates a terminal building that embodies both the regional characteristics of Shanxi and the appearance of the new era.

The external form of the terminal building adopts a "big roof" design, achieving a harmonious unity of spatial function, structure, and appearance. The facade reflects the solemn temperament of historic city, while the stepped internal design creates a unique spatial experience. Herringbone metal trusses support the "big roof," extending to the ground on both sides. The long-sloped roof on the landside provides rain shelter for the driveway, and the short-sloped roof on the airside extends the sightlines of waiting passengers and facilitates the layout of boarding bridges, ensuring a harmonious blend of architectural form and spatial function. (Figs. 5-14, 5-15)

In summary, the design of the T2 terminal at Datong Yungang International Airport in Shanxi not only inherits regional culture, but also highlights the characteristics of the times through the abstract extraction of traditional roof elements, reflecting the perfect integration of architecture and culture.

Figure 5-14 Front facade of Datong Yungang International Airport

Figure 5-15 Landside interior of Datong Yungang International Airport

Case Study: Shennongjia Airport in Hubei, China
External styling design inspired by Myth of Shen Nong

The terminal building of Shennongjia Airport in Hubei, China completed in 2013, is a small regional airport covering an area of 4,000 square meters with an annual passenger throughput of 250,000. The design concept is deeply influenced by the local Shen Nong culture, showcasing unique regional characteristics and cultural connotations through modern architectural techniques. Inspired by the myth of "Shen Nong building houses with wooden frames," the abstract elements of triangular structures are applied to the roof design, imparting spiritual symbolism. The wooden triangular folded metal roof harmonizes with the surrounding undulating mountains while also reflecting the mysterious charm of legendary stories. The building facade adopts a symmetrical design with triangular shapes, creating strong visual tension that leaves a lasting impression on arriving passengers and allows them to experience the cultural significance of wooden structures created by Shen Nong during their waiting for boarding, evoking cultural resonance. Additionally, the natural wood color of the roof extends into the terminal's interior space, ensuring that the "wooden structure" design concept is consistently reflected in the integrated indoor and outdoor design. (Figs. 5-16, 5-17)

Figure 5-16 Front facade of Shennongjia Airport terminal

Figure 5-17 Detailed roof structure of Shennongjia Airport terminal

2. Spatial Metaphor of Regional Culture

Architecture, as a material carrier of culture, plays a crucial role in conveying history and tradition. As the aerial gateway to a region, airport terminals must prioritize the inheritance and expression of culture. Each country and region in the world possesses unique human history, myths, legends, and traditional architectural styles. By abstracting these cultural images and integrating their distinct spatial characteristics and cultural connotations into terminal design, architects can subtly and implicitly incorporate these elements into the spatial design of airport terminals.

Spatial metaphors of regional culture not only provide a profound expression of the local temperament, but also achieve an organic integration with the airport's local environment. This design approach imbues the spatial image with subjective color, sparking people's imagination about the space's origin, and injecting local spirit and cultural significance into the space.

Case Study: Terminal 2 of Mactan-Cebu International Airport in the Philippines
Architectural forms rooted in local cultural context

Mactan-Cebu International Airport, the second largest airport in the Philippines, constructed Terminal 2 adjacent to the original Terminal 1, serving over four million person-times annually. Although the terminal is larger than those of typical small- and medium-sized airports, its design approach rooted in traditional cultural context offers valuable insights. Cebu is a popular holiday destination in the Philippines, and the new terminal serves as an important gateway for serves passengers. This terminal not only fulfills its transportation function, but also showcases local

Figure 5-18 Perspective view of Terminal 2 of Mactan-Cebu International Airport

culture. Its design draws inspiration from the locally tropical dwelling characteristic, featuring high slope roofs and low eaves. These design elements effectively adapt to the hot and rainy climate, creating a more comfortable indoor environment. (Fig. 5-18)

The terminal's first level is constructed with reinforced concrete, while the second level employs traditional wooden structures, making it light and easy to build. The massive wooden beams of the roof are hinged to the reinforced concrete base of the first level. The lightweight wooden structure and joint design that accommodates some degree of misalignment enhance the building's ability to withstand typhoons and earthquakes. Continuous wooden arches add warmth to the airport, distinguishing the Mactan-Cebu International Airport terminal from other transportation buildings and creating a welcoming atmosphere that allows passengers to feel the vacation vibe upon arrival. (Fig. 5-19)

Figure 5-19 Interior space of Terminal 2 of Mactan-Cebu International Airport

Summary: Innovative design concepts based on humanistic attributes yield two specific creative strategies:

Abstract extraction of cultural symbols: This strategy involves extracting and translating cultural symbols with specific meanings into architectural forms, colors, and textures. For example, the design of Shennongjia Airport terminal abstracts traditional wooden frame construction techniques into triangular folded plate elements.

Spatial metaphor of regional culture: This approach involves extracting cultural connotations and spiritual values and expressing them metaphorically in spatial design. This method is more subtle and nuanced, achieving a deeper integration of local culture into the airport terminal's design.

5.2.1.3 Summary of External Form Innovation

This section summarizes two creative ideas that influence the external form of regional airport terminals. The first is based on natural elements, namely drawing on topography, adapting to natural climates, and adopting bionic abstract forms to create exterior designs of airport terminals that harmoniously coexist and engage in dialogue with nature. The second stems from humanistic attributes, utilizing the abstract expression of cultural symbols and the spatial metaphor of regional culture to create designs with regional cultural characteristics. These creative ideas are influenced by both the subject and the object of design, showing rich flexibility. In design practice, it should not be limited to the two design ideas mentioned above.

The external form cannot exist in isolation from spatial function. In addition to focusing on the exterior design of airport terminals, attention to internal space and passenger experience is also the key content of innovation. The following section will focus on the innovative design methods of the internal experience of the terminal buildings.

5.2.2 Innovative Design of Internal Experience

The internal spatial experience of an airport terminal is mainly influenced by basic functions such as ticketing, security checks, waiting for boarding, and baggage claim. With the development of society and the change of people's lifestyle, terminals that only meet the basic functions have been unable to meet the needs of the times, and the design focus of terminals has gradually shifted from functionality to aesthetic, cultural and spatial experience. This transformation requires architects to focus on integrating new functions and technologies into spatial and experiential design, delving into regional characteristics, pursuing humanization, and leveraging technological advancements based on the foundations of terminal construction of old eras.

Through case studies, this section explores three innovative design techniques for terminal internal experience: new function integration, green ecology incorporation, and cultural experience enhancement, with a view to providing a reference for the innovation of the internal spatial experience of the terminal building and stimulating more possibilities.

5.2.2.1 New Function Introduction

The main functional space of the airport terminal building generally includes the departure hall, check-in and ticketing hall, security checkpoint, boarding lounge, VIP lounge, arrival corridor, baggage claim hall, arrival hall, and baggage handling area. Simply arranging these functional areas under the roof of the airport will result in a monotonous spatial experience and lack of creativity.

This section summarizes the possible applications of several new types of functional spaces in terminal buildings through case studies. For example, the introduction of landscape space allows users to watch aircraft takeoffs and landings from observation decks in the terminal; the introduction of art and exhibition space allows people to explore aviation-related knowledge in airport museums or enjoy experiences similar to strolling through an art gallery ramp in the airport; the introduction of commercial and leisure functions changes the hectic atmosphere of typical airports so that travelers can feel relaxed in airports; the integration of communication and interaction functions allows individuals to spend heartwarming moments with family and friends in specific farewell spaces before takeoff. These new functions, while meeting basic requirements, incorporate more human-centered design elements, greatly enhancing users' travel experience.

1. Scenic Observation

Airport terminals have a unique view of the takeoff and landing of airplanes. Typically, only passengers waiting in the boarding area can see this spectacle through the terminal's windows. However, some architects provide opportunities for non-passengers to also experience this phenomenon by integrating innovative design concepts. Designers have created dedicated outdoor observation decks, offering people an excellent vantage point to admire and photograph airplanes. These observation decks are also integrated with casual dining areas, creating a unique dining experience.

Buildings should not be barriers that restrict people but rather mediums that guide people towards enhanced experiences. While meeting security requirements, airport terminals should strive to be as open as possible, conveying their aviation-related features and advantages to a broader range of users.

Case Study: Florianópolis-Hercílio Luz International Airport in Brazil
Independent observation deck separated from the main building

This airport terminal building has a total area of approximately 45,000 square meters and an annual passenger volume of around 8 million person-times, serving the Brazilian capital and a popular tourist city nearby - Balneário Camboriú. During the best sunny days in summer, This city attracts a large number of tourists from Brazil, Argentina, Uruguay, Chile and Paraguay to its beautiful beaches. Many of them come to enjoy observing and photographing airplanes.

To meet these needs, Biselli + Katchborian Arquitetos Architects, who designed the terminal, installed a 600-square-meter observation deck at the top of the right side of the terminal. The entrance to the observation deck is located next to the outdoor bus stop on the ground floor, and visitors can quickly reach it through a separate elevator in the core-tube without entering the interior of the terminal building. Here, people can watch activities on the runway and apron, and also rent the terrace for various events and gatherings. This innovative observation deck has become a place that inspires and explores the spectacular beauty of the aviation world in the new terminal building. (Figs. 5-20, 5-21)

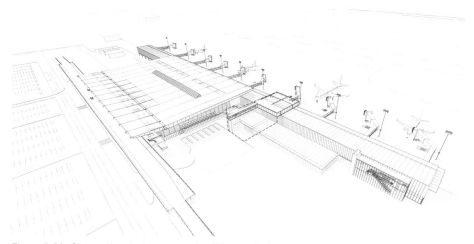

Figure 5-20 Observation deck on one side of the terminal

Figure 5-21 Observation deck and the users on it

2. Art and Exhibition

When people think of art and exhibition spaces, they usually imagine art galleries or museums. However, combing these elements with airports can offer travelers a whole new experience. The following two cases illustrate this concept: one features an "Airport Museum" located between two terminal buildings, and the other utilizes the common ramp design seen in art galleries. Both cases attempt to introduce the experience of wandering through space into a terminal that aims for efficiency.

These two cases are highly valuable for research because they add educational and entertainment experiences to the airport's transportation and commercial characteristics. This attempt to overlay different types of functions within the terminal building is highly inspiring for innovation in airport terminal design.

Case Study: Chubu Centrair International Airport in Nagoya, Japan
Airport museum with educational and commercial functions

The Airport Museum is an independent structure located between the first and second terminals of Chubu Centrair International Airport in Nagoya, Japan. The buildings are connected by automated pedestrian walkways, and it takes approximately 5 to 10 minutes to walk from either terminal to the museum. (Fig. 5-22)

In the exhibition area on the first floor, visitors can take a close-up observation of the Boeing 787 and learn about the mechanical components of aircraft and aviation-related work through actual exhibits, display boards, and films. In the children's area, kids can play in amusement facilities near the aircraft and enjoy a unique perspective of airplane takeoffs and landings. The restaurant and store on the second floor allow visitors to enjoy authentic Seattle cuisine and purchase

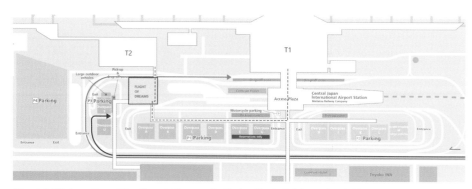

Figure 5-22 Location of the Airport Museum in Chubu Centrair International Airport

Figure 5-23 Physical exhibits of aircraft

Figure 5-24 Display zone of the Airport Museum

specialty items at the Boeing official store. (Figs. 5-23, 5-24)

Although the Airport Museum is not located inside the terminal, its innovative use of functions has significant reference value. The diverse functions of the terminal help enrich the space types and improve the travel experience of passengers.

Case Study: Concourse D at Seattle-Tacoma International Airport in Washington, USA

Airport annex building with ramp roaming experience

The concourse is a two-story airport annex building designed to alleviate congestion in the main terminal. Serving as a passageway between the main terminal and the apron, it includes 6 passenger boarding gates, concession stands, and a children's play area. Departing passengers complete security check and verification procedures in the main terminal before entering the second floor of the hall via a skybridge. They are then greeted by a prominent ramp that occupies a large portion of the space, guiding passengers down to the departure hall on the first floor.

Ramps, known for their gradual, long, and winding characteristics, are often used in cultural buildings such as art galleries and museums, but rarely seen in transportation structures like airports that prioritize efficient traffic flow. However, the architects experimented with ramps in this small airport annex building, aiming to slow down users' pace and allow them to appreciate their surroundings as they gradually descend. These views include the intricate interior architecture, moving or standing passengers, and the cityscape outside the windows. When passengers have enough time, the ramp offers them a roaming and observational experience during their boarding processes. (Figs. 5-25, 5-26)

INNOVATION DESIGN INNOVATION OF REGIONAL AIRPORT TERMINAL BUILDING

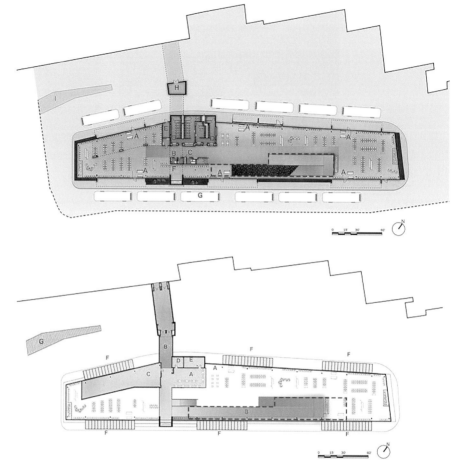

Figure 5-25 The ramp on the first and second floors in the airport annex building

Figure 5-26　The ramp and the wandering travelers

3. Commercial and Recreational Functions

Terminal buildings are usually lined with shopping and food outlets, with the main purpose of generating economic benefits, and users can only stay in specific spaces. However, the following case study presents a more leisure-oriented spatial concept: the integration of a "brand center" that combines commercial and leisure functions with local characteristics. The open space is defined by furniture rather than separate rooms. The use of appropriate materials and the integration of interior and exterior design create a more relaxing atmosphere.

While this approach may seem to contradict commercial profitability, it is possible that people may better appreciate and understand the brand concepts in a more relaxing atmosphere, and may be more willing to make purchases in such a setting.

Case Study: Kangaroo Island Airport in Australia
Introducing a brand center to create relaxing atmosphere

The airport is located on Kangaroo Island, a famous tourist destination in southern Australia, with a building area of approximately 2,100 square meters and an annual passenger throughput of around 50,000 person-times. As it is a small regional tourist airport, the architects focused on how to embody the spirit of Kangaroo Island within a limited space, promoting interaction between people and nature, showcasing the authenticity and diversity of the island's people, and providing genuine experiences.

To achieve this, the architects introduced a "Kangaroo Island Airport Brand Center" within the terminal building, which includes a kitchen, a bar, an art gallery, and plenty of comfortable seating areas. The brand center faces a spacious lawn bathed in sunlight, with indoor wood paneling seamlessly transitioning to outdoor wooden decking. On warm days, the windows can be fully opened to bring the island's unique natural beauty inside the building. (Figs. 5-27 ~ 5-29)

Arriving passengers can get their first impressions of Kangaroo Island by visiting a gallery showcasing photographs of the island's natural habitat while waiting for a taxi to their destination. Departing passengers, while waiting to board their flights, can savor a cup of coffee, find a favorite seat, and once again relish the charm of the island. This airport structure successfully conveys the spirit of Kangaroo Island, optimizes the passenger experience, listens to the needs of its users, and showcases its unique charm.

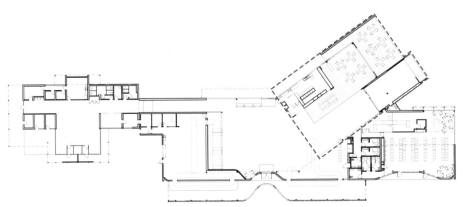

Figure 5-27 Plan of the first floor of Kangaroo Island Airport terminal, showing the location of the Brand Center

Figure 5-28 Lawn facing the Brand Center

Figure 5-29 Blurred indoor and outdoor leisure space

4. Interaction and Communication

This section focuses on the unique farewell scenes within the terminal building. In many countries and regions, there is a strong emphasis on the tradition of friends and family accompanying travelers until the very last moment before boarding, leading to the creation of dedicated farewell and greeting spaces within airport terminals. The following two European case studies illustrate how these farewell areas have become highlights within small- and medium-sized airport terminals.

The purpose of studying these cases is to inspire airport designers to consider interpersonal relationships as a key aspect of spatial design. By studying people's behavioral patterns, designers can create truly personalized and functional spaces that cater to the needs of individuals.

Case Study: Kaunas Airport in Lithuania

Top-floor farewell space for friends and family overlooking security checkpoint and boarding lounge

Kaunas is located in the southern part of Lithuania and is the historical capital of the country, attracting millions of European history and culture enthusiasts annually. The airport has a total area of 7,378 square meters and serves passengers of over 1 million person-times per year. The ground floor serves as the arrival level, while the second and third floors function as the departure level. The terminal building was designed in a square shape with a shorter entrance side and a longer side, allowing for future expansion on both sides. As a result, the area of the front arrival/departure hall on the second floor is constrained by the security checkpoint, making the space for passengers and friends and family farewells appear cramped. (Fig. 5-30)

To address this issue, the architects created a long rectangular space on the third floor, providing travelers and their loved ones with a place to spend time together before the journey (Fig.5-31). The access to this farewell space is easy, with an open elevator that goes straight up to the third floor directly opposite the escalator platform connecting the first and second floors. The long space has plenty of lounge seating at the front, a cafe at the back, and a view of airplanes landing and taking off through the windows at the end. In this farewell space, people can engage in face-to-face conversations and interactions, while friends and family members can also look down and watch as travelers move from the security checkpoint into the boarding lounge. (Fig. 5-32)

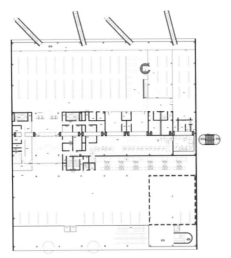 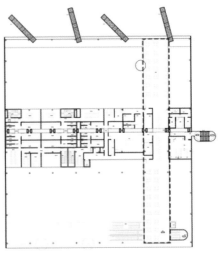

Figure 5-30 Cramped departure hall on the second floor

Figure 5-31 Farewell space for family members and friends on the third floor

Figure 5-32 Resting seats in the farewell space on the third floor and a downward view of the security checkpoint

Case Study: Nuuk Airport in Greenland

Farewell space for friends and family members overlooking the boarding lounge

Nuuk Airport is located in the capital of Greenland, with a building area of 9,200 square meters. Due to its remote location and limited transportation options, local residents often have to travel for extended periods when flying out of Nuuk. As a result, there is a tradition of friends and family members coming to the airport to say goodbye until the very last moment before departure.

The architectural design team ZESO, guided by the principle of "people-oriented design," created a farewell space for Nuuk Airport. Visitors enter the farewell hall on the second floor through a gate, where they can find a café, seating areas, and steps for lingering. As boarding time approaches, travelers proceed through security on one side and go downstairs to the departure lounge via an escalator. A large floor-to-ceiling glass separates the farewell area on the second floor from the departure lounge on the first floor, allowing friends and family members to watch as travelers board the plane. They can bid greetings and farewells in a tranquil and natural environment, enjoying a warm experience provided by the airport. (Figs. 5-33, 5-34)

Figure 5-33 Farewell space for family members and friends in Nuuk Airport terminal

Figure 5-34 Rendering of the farewell scene in Nuuk Airport terminal

Summary: Through the cases analyzed in this section, it is evident that small- and medium-sized airports have various options for functional space innovation, each with unique spatial forms.

Scenic observation: Observation decks or spaces do not need to be overly large but should be widely applicable, which makes them achievable in almost all terminal buildings. Being combined with surrounding landscape resources, the scenic observation facilities allow visitors to enjoy natural scenery and the spectacle of airplanes taking off and landing, adding a major highlight to the terminal building.

Art and exhibition: To make these bring distinctive experience in the terminal building, it is not enough to simply set up exhibition spaces. Designers need to realize the concept of airport museums/art galleries by setting up specific spatial elements. For example, utilizing ramps to create a roaming viewing experience or using double-height space combined with mezzanines to enable interaction between users and large exhibits. Furthermore, in terms of circulation design, art and exhibition areas should be placed on other major circulation routes outside of the boarding lounge, allowing both travelers and their loved ones to participate.

Commercial and recreational functions: It is suggested that airport owners and designers explore local characteristics, establish airport brand positioning, and set up experiential commercial spaces, which may more effectively achieve commercial results.

Interaction and communication: Although the concept of farewell spaces for family and friends in airports mentioned above originates from foreign cases, this humanizing experience is equally applicable to Chinese airports. Typically, family members and friends sending someone off at the airport can only stay in a crowded hall filled with ticket counters and limited seating, waving goodbye in front of the security checkpoint, which makes the parting seem a bit hasty. It is recommended that mezzanines be set up in the hall, allowing family members and friends to watch the travelers pass through security check, or establishing visual connections between the hall and the boarding lounge, providing users with a warmer farewell experience.

Functional innovation is not limited to the four categories discussed in this section. Designers should explore more possibilities from the local natural environment and cultural characteristics, continuously driving innovation in airport design to enhance the travel experiences of global passengers.

5.2.2.2 Nature and Ecology Incorporation

Modern airport terminals place a strong emphasis on green ecological design, encompassing aspects such as the surrounding environment, natural landscape, ecology, climate, lighting, and ventilation. This section focuses on how to integrate green ecological concepts into the interior space of airport terminals. Through domestic and international cases, we have extracted the design techniques that can be utilized, including oppositive scenery introduction, integration with nature, and light and shadow creation. These skillful spatial design techniques can encourage interaction between passengers and nature, bringing innovative experiences to interior space.

1. Oppositive Scenery Introduction

Airport terminal design can introduce the surrounding environment and natural landscapes through the use of sightlines, and the design techniques such as creating oppositive sceneries, enframed sceneries, and negative space can be skillfully used to achieve visual interaction between interior space and the external environment. This design approach can create unique visual relationships and viewing experiences, allowing passengers to have dialogue with nature and enjoy the physical and mental pleasure brought by the natural environment. For example, Gibraltar Airport utilizes the unique surrounding landscape to extend the beautiful rock scenery into the interior space of the terminal building.

Case Study: Gibraltar Airport
Airport facing natural rock scenery

The new terminal building at Gibraltar Airport covers an area of 20,000 square meters and is designed to accommodate an annual passenger throughput of 1 million person-times. The airport's geographical location is unique and complex, with strict environmental restrictions on all sides. Due to the airport's proximity to a rocky landscape, the designers cleverly utilized this landscape resource by orienting the airside of the terminal building towards the magnificent rock scenery. They designed a spacious and bright airside rooftop terrace as an extension of the boarding lounge for passengers to relax and enjoy the rock scenery, creating an environment-sensitive building that interacts with nature. (Fig. 5-35)

The roof design of the terminal building draws inspiration from sailboats on the sea, with wide eaves providing shading, while the glass curtain wall meets the natural lighting needs of the terminal building, ensuring clear visibility and providing expansive scenic views. Passengers waiting indoors can admire the picturesque rock scenery, and they can also step outside to bathe in the sunlight, interact, and communicate, maximizing their enjoyment of the natural landscape. The design technique of landscape permeation gives Gibraltar Airport a unique viewing experience and regional characteristics. (Fig. 5-36)

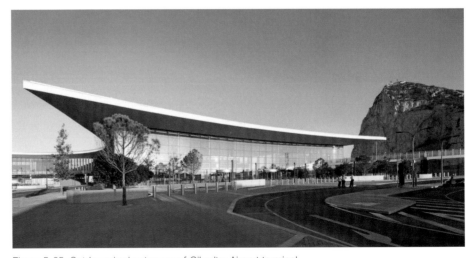

Figure 5-35 Outdoor viewing terrace of Gibraltar Airport terminal

Figure 5-36 Gibraltar Airport's landscape design features outdoor oppositive scenery

2. Integrating with Nature

Green ecological concept plays a significant role in airport terminal design, with designers typically incorporating green and energy-saving design elements through external form, eco-friendly materials, and energy-efficient ventilation. However, few designers consider incorporating natural design methods from the perspective of functional usage and spatial patterns in airport terminal design, but these methods can provide an innovative direction and approach for terminal design.

Integrating ecological functions into spatial design under specific climatic conditions can bring about innovation in the spatial patterns of airport terminals. Introducing courtyard space with natural scenery not only effectively regulates the indoor microclimate, but also provides passengers with a unique garden-like experience. In foreign countries, small- and medium-sized airport terminals commonly adopt spatial design methods that integrate ecological functions, while Chinese airport terminals are actively exploring the possibilities of achieving spatial innovation through "integrating with nature." The following two cases featuring natural courtyard integration provides new ideas for innovative design.

Case Study: Ramon International Airport in Israel
A touch of green in the desert

Ramon International Airport, located near a harbor city Eilat in southern Israel, covers a total area of 45,000 square meters and is designed to accommodate an annual passenger throughput of 2.5 million person-times, featuring a typical one-and-a-half-story terminal building. Surrounded by natural desert landscapes, Ramon International Airport experiences hot and arid climates. Drawing inspiration from the mushroom-shaped rock layers in the national park near the airport, the designers mimicked the process of rocks being shaped by wind and water erosion. They employed a rugged and minimalist design language that harmonizes with the surrounding environment, with the sturdy volume of the terminal building helping to withstand the impact of harsh external natural conditions.

The exterior of the terminal building is continuous and cohesive. The interior space undergoes excavation, with a winding landscape strip dividing the building volume into two parts, segregating the airside and landside functions. A central courtyard is embedded in the main space, bringing the natural desert landscape into the interior of the building and effectively solving the problem of natural lighting. Departure and arrival halls are arranged around the central courtyard, allowing passengers to enjoy natural scenery in various functional areas of the airport. Additionally, the airport preserved seeds of local plants on the site in greenhouses for cultivation during

Figure 5-37 Aerial view of Ramon International Airport

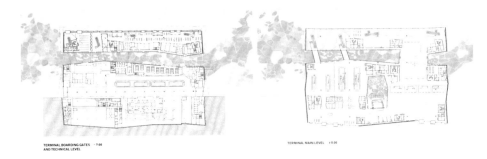

Figure 5-38 Site plan of Ramon International Airport

construction and replanting them back to original locations after completion, transforming the airport courtyard landscape into a desert oasis and adding a green delight to the passenger experience. The innovative interior design makes Ramon International Airport the first civil airport following green development idea in Israel. (Figs. 5-37, 5-38)

Case Study: Shangrao Sanqingshan Airport in Jiangxi, China
Creating poetic space with light atriums

Shangrao Sanqingshan Airport is located in Jiangxi Province, China and officially commenced operations in 2017. The terminal building covers an area of 10,496 square meters and is designed to accommodate an annual passenger throughput of 500,000 person-times, featuring a typical one-and-a-half-story terminal building. The airport is situated among the mountainous area in eastern Jiangxi, seeing distinct subtropical monsoon climate with abundant rainfall and plentiful water resources.

The terminal building adopts the design concept of "Rain in the Remote Mountain." The external form echoes the surrounding natural environment, and the undulating lines of the roof seem to show the style of natural hills. The interior space incorporates three circular atriums for natural lighting as a connection between people and nature, allowing passengers to experience the changing scenery of four seasons in indoor space. The departure hall on the ground floor with light atriums serves as the first impressive spot upon entering the terminal building. The boarding lounge on the second floor is positioned next to the atriums, enabling passengers to enjoy natural lighting from the courtyard and gaze at the natural beauty while waiting, which brings passengers an experience with unique regional characteristics. The interior design of the terminal building uses abstract methods to connect with the atriums, with the ceiling forming concentric light bands symbolizing the ripples on a water surface, creating a poetic space filled with natural sentiment. (Figs. 5-39, 5-40)

Figure 5-39 Exterior view of the terminal of Shangrao Sanqingshan Airport

Figure 5-40 Interior space of Shangrao Sanqingshan Airport terminal

3. Light and Shadow Creation

Light and shadow play a crucial role in airport terminal design, requiring designers to meticulously control these effects. Beyond fulfilling basic natural lighting need, the creative use of light can enhance the spatial experience. Terminals can employ design techniques to introduce natural light into interior spaces, with subtle light and shadow effects creating varied atmospheres. The interplay between architecture and light can provide passengers with a unique and rich experience. For example, the Pamplona Airport in Spain emphasizes the rhythmic beauty of its roof design and natural lighting, while the Florianópolis International Airport in Brazil uses sloping walls to direct light, creating memorable and innovative spaces. Here, the author will detail these two case studies.

Case Study: Pamplona Airport in Spain
Art gallery-style natural lighting

Pamplona Airport, located in northeastern Spain, is a medium-sized civil airport for the city of Pamplona. It features a one-and-a-half-story terminal building. The interior is divided into functional areas by five courtyards, with a 12-meter-high central concourse connecting the departure, arrival, security check, and waiting areas, making it the most spacious and well-lit public area in the terminal.

The terminal makes clever use of natural lighting through carefully designed roof skylights, providing a large amount of natural light and creating comfortable spaces for passengers. The skylights use proportionally arranged roof units in alternating horizontal and vertical rows, creating a rhythmic and aesthetic natural light effect, adding layers of light and shadow. To enhance natural lighting effect, each metal component is meticulously shaped like a shuttle, and paired with warm-colored roof materials, creating a warm and comfortable natural light ambiance. The soft light falling from the skylights brings about an art gallery-style light and shadow experience. This design enables the terminal to rely on natural light for indoor illumination, with courtyard spaces contributing to ventilation, lighting, and microclimate regulation, making the terminal an exemplar of low-carbon design. (Figs. 5-41, 5-42)

Figure 5-41　Facade of the terminal building of Pamplona Airport

Figure 5-42　Interior space of Pamplona Airport terminal

Case Study: Florianópolis International Airport in Brazil
Utilizing slanted walls to introduce natural light

Florianópolis International Airport features ten boarding bridges and adopts a classic two-story terminal building design. The T-shaped terminal structure consists of administrative areas and the arrival and departure halls, with the departure hall being a double-height, column-free, transparent space (Fig. 5-43).

The architects carefully considered the use of natural light in the layout and decoration of the terminal, particularly in the interior of the departure hall. The innovative design of slanted wooden walls not only cleverly conceals air conditioning pipelines, but also integrates with the folded design of the roof skylights to guide natural light deeply into the hall, offering passengers a comfortable experience of natural lighting. This design not only enhances the visual appeal of the terminal, but also creates a unique atmosphere that interacts with light and shadow. Especially with skylights positioned above the check-in counters and garden courtyards, sunlight filtering through the walls creates a vivid interaction of light and shadow, providing passengers with a dreamlike experience. (Fig. 5-44)

The architectural aesthetics of this terminal, engaging in a dialogue with light and shadow, not only enriches high-quality visual experience for passengers, but also seamlessly integrates time and art, adding a unique charm to travel.

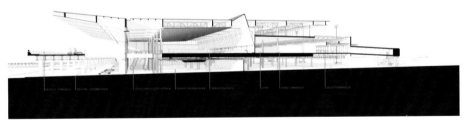

Figure 5-43 Cross-sectional diagram of Florianópolis International Airport

Figure 5-44 Light and shadow effects in the interior of Florianópolis International Airport

Summary: This section summarizes three design strategies of "green ecology incorporation" in exemplary airport terminals: oppositive scenery introduction, integration with nature, and light and shadow creation. These strategies are achieved through the interaction of natural elements with internal spaces and should be applied judiciously based on specific circumstances. For oppositive scenery introduction, leveraging the surrounding natural landscape through landscape design, framing design, and negative space treatment can enhance visual continuity between indoor and outdoor spaces. Integration with nature involves integrating natural elements like atrium gardens or light wells into indoor spaces to immerse passengers in natural environments. Regarding light and shadow creation, besides glass curtain walls, skylights play a crucial role in achieving natural lighting and creating unique lighting effects. Architects can ingeniously design skylights to harmonize with indoor materials and forms, thereby enhancing space quality and comfort.

Skillfully integrating natural elements into airport terminal interiors not only enhances passenger experience, but also injects vitality and innovation into spatial patterns. The application of natural elements such as plants and sunlight in spatial design merits further exploration and represents a pivotal direction for future innovations in airport terminal experiences.

5.2.2.3 Cultural Experience Enhancement

Culture represents the shared values and standards of a social group, and incorporating cultural features into architecture aims to showcase these characteristics to users, giving them a direct impression of a specific culture. Different regions possess their own distinct cultural traits, adding diversity to architectural experience. The following case study illustrates how terminal design effectively integrates cultural identity.

Case Study: Terminal 1 of Baku Heydar Aliyev International Airport in Azerbaijan
Cultural experience of Azerbaijan's Silk Road silkworm cocoon imagery

Located in Baku, the capital of Azerbaijan, this airport terminal spans approximately 65,000 square meters and has an annual passenger throughput of around 6 million person-times. Azerbaijan Airlines, the airport's owner, wanted the design to reflect Azerbaijan's culture, values, and national image. Seyhan Özdemir, co-founder of the designer of this project Autoban Studio, explained: "Airports are quickly becoming destinations in the tourism industry. They are the first faces that welcome travelers from all over the world, where you can get the first impression of their culture."

In line with this vision, Autoban designed custom wooden "cocoons" for the airport interior. These cocoons reflect Azerbaijan's historical significance on the Silk Road during the 18th and 19th centuries and symbolizes the wealth generated from silkworm breeding and silk trade. The second floor of the boarding lounge features 16 wooden cocoons of similar but not identical shapes, serving various purposes such as café, bar, children's play area, spa and beauty salon, music bookstore, and luggage storage facilities. Inspired by the hospitality of the Azerbaijani people, the designers arranged the cocoons flexibly, creating a sense of wonder and discovery for passengers, with numerous opportunities for encounters and exploration. The clusters of wooden cocoons scale down the vast airport space to a more human-centric size, conveying a "human-oriented" spirit to passengers. (Figs. 5-45, 5-46)

Figure 5-45 Wooden cocoons on the top floor of Baku Heydar Aliyev International Airport

Figure 5-46 Encounter space between wooden cocoons

Summary: Under the influence of modern architectural thinking, there are relatively few successful cases of cultural experiences in airport spaces. When incorporating cultural elements into architectural space, the key lies in high refinement and abstraction, allowing people to naturally feel the cultural atmosphere through transformation and integration. It is crucial to avoid the direct collage of cultural elements on spatial components to prevent visual confusion and maintain efficiency.

5.2.2.4 Summary of Internal Experience Innovation

Through the study of excellent cases at home and abroad, this section summarizes three innovative design approaches that influence the interior design of airport terminals, highlighting the integration of new functions, green ecological elements, and cultural experience.

The innovative idea of new function introduction involves scenic observation, art and exhibition, commercial and recreational functions, interaction and communication, etc., which can each bring unique and innovative experiences. This approach is widely applicable, and the combination of terminal building and composite functions can burst new spatial vitality, which is also the key research direction of modern terminal building space innovation.

The innovative idea of nature and ecology incorporation utilizes design techniques such as oppositive scenery introduction, integration with nature, and light and shadow creation to create unique experience related to natural ecology. This approach is highly applicable, especially since natural ecology is a significant topic in various modern architectural designs. Integrating airport terminals with natural experience is becoming an essential aspect of future airport terminal designs.

The innovative idea of cultural experience enhancement involves spatializing and experiencing regional characteristics to generate unique cultural experiences specific to the locality. This approach requires careful application, considering that airports are modern architectural structures rather than traditional buildings. It is crucial to avoid directly collaging traditional elements into the design. However, if traditional culture and modern airport space and functions can be skillfully combined, it can become an innovative highlight of the airport terminal.

The research and summary of these excellent cases provide a framework, but only through deeper and broader exploration of nature and humanities can designers continuously expand this framework and create updated and unique airport terminal spatial experiences.

5.2.3 Innovative Design in Detailed Construction

Colors, materials, and construction details are important components of airport terminal design and key aspects that influence the effectiveness of spatial design. By creating details such as color schemes, materials, and construction elements in airport terminals, the most direct spatial experience and aesthetic value can be provided to passengers, helping to express design concepts and enhance passengers' rich and detailed spatial experiences.

By collecting excellent Chinese and international case studies in recent years, innovative design methods for detailed construction can be roughly categorized into the following three points: the use of colors, the use of materials, and the use of construction techniques. From the perspectives of visual aesthetics and spatial perception, innovative design methods for detailed construction in airport terminals will be discussed in the following sections.

5.2.3.1 Innovative Use of Colors

The selection and use of colors is crucial to the overall effect and spatial experience of a terminal building, which can influence people's perception and experience of the space, as well as convey the design concept, guide the line of sight, shape the mood and express cultural connotations. Skillful use of color in airport terminal design can present different atmospheres and emotions, thus creating a unique and rich spatial experience for users. Therefore, in the process of terminal design, colors should be regarded as an important design element, which needs to be carefully considered and designed.

The following section summarizes color design techniques commonly used in the design of airport terminals, and analyzes them from four aspects: visual enhancement, perceptual experience, spatial guidance, and cultural representation.

1. Using Colors to Enhance Visual Effects of Space

As colors can attract passengers' attention and enhance visual effects, airport terminals can utilize colors to highlight key spaces and use bright colors to emphasize structural components of the airport. This makes the airport's features stand out at a glance while also enlivening the spatial atmosphere.

Case Study: Madrid Airport in Spain

Conveying enthusiasm and vitality through supportive structures

Although the size of Madrid Airport far exceeds that of small- and medium-sized airport terminals, its use of colors is worthy of consideration and reference. The airport features a dynamic array of Y-shaped support columns that hold up the undulating floating roof. These columns have become the most prominent spatial elements under the unique roof. Most of them are painted in red and yellow, while a few feature a rainbow gradient color. In addition to highlighting the structural features of the airport, the design also conveys the passionate, exuberant, and vibrant characteristics of the city of Madrid to arriving passengers. (Fig. 5-47)

Figure 5-47 Madrid Airport terminal: Yellow color is mainly used in the interior space, and red color is used in the airside interface

2. Emphasizing Spatial Perception through Colors

Different colors can evoke different spatial experiences and impact human psychological behavior. For example, an airport that extensively uses wood, stone, or wood-grain new materials will create a warm, safe, and comfortable atmosphere throughout the terminal. In contrast, an industrial color scheme of black, white, and gray conveys a modern sense of efficiency and simplicity.

Case Study: Helsinki Airport Terminal in Finland
Green architecture with wooden ceiling structure

The Helsinki Airport utilizes a wooden ceiling design. The vertical stacking of wood enhances the spatial perception of the design, and the wavy form presents a unique aesthetic appeal even from a distance. The timber used for the ceiling is Finnish spruce, which is not only resistant to decay and deformation, but also ensures the stability of the structure. The unique soft colors and natural textures showcase simple and primitive beauty, revealing Finland's unique mild temperament. (Fig. 5-48)

The security checkpoint within the terminal is adorned with a light blue color, which stands out prominently within the warm yellow space, visible even from a distance. This design allows passengers entering the departure hall to immediately see the blue security checkpoint in the distance. In addition to enhancing legibility, the blue color also plays a role in soothing and calming individuals during the security check. (Fig. 5-49)

Figure 5-48 Helsinki Airport terminal with cascading wooden ceiling extending from outdoors to indoors

Figure 5-49 Helsinki Airport terminal with warm-toned wooden ceiling connecting to the blue security checkpoint

Case Study: New Terminal of Quebec Chibougamau-Chapais Airport in Canada
A warm stop in high-latitude regions

Chibougamau-Chapais Airport is located in the high-latitude region of Quebec, Canada. The vast land and cold climate of this region set the tone for the construction of the new terminal. The design of the new terminal highlights the characteristics of the northern forest, incorporating locally produced wood. The terminal is composed of a glazed concourse and two white, low-volume structures on either side. The roof of the waiting area and all four sides extensively use laminated wood, creating a warm and inviting spatial experience. The white exterior volumes complement the surrounding environment, while the interior, filled with light yellow color of wood, exudes a warm atmosphere enveloped by high-transparency glass, welcoming arriving passengers with a sense of warmth. (Figs. 5-50, 5-51)

Figure 5-50 Chibougamau-Chapais Airport terminal, white volumes blending with the surroundings

Figure 5-51 Interior space of Chibougamau-Chapais Airport with extensive use of wood

3. Using Colors to Enhance Visual Guidance

Vibrant colors play a significant role in creating striking visual effects, and the bold use of colors in airport terminals can leave passengers deeper impression. Through the use of different colors, it is possible to differentiate spaces of various functions and guide passengers effectively. While airport spaces mostly adopt gray, white, and blue tones to reflect efficiency, simplicity, and international characteristics, some airports utilize the psychological impact of colors and daringly incorporate unconventional colors to create indoor spaces that convey unique feelings, enhancing passengers' impression and experience of the airport.

Case Study: Connecting Passageways of Narita Airport Terminal 3 in Japan
Assigning colors to different functional spaces

The spatial design of Narita Airport is based on efficiency and simplicity, using ground color to delineate the departure and arrival passenger flows, as well as the passage and waiting areas in the baggage claim hall. The blue color represents efficiency and calmness, guiding the departure flow, while the dark red color brings a hint of passion and impulsiveness, guiding the arrival flow. The two colors complement each other to guide the flow of passengers during peak hours, and the different area ratios of the two colors imply the number and proportion of different flows of people. (Fig. 5-52) Additionally, low-saturation green and blue are used in the rest areas to soothe the anxiety of waiting passengers (Fig. 5-53). The bold use of colors adds more vitality and humanistic care to the originally efficient terminal.

Figure 5-52 Connecting passageways at Narita Airport Terminal 3 differentiated by red and blue colors to indicate different functional flows

Figure 5-53 Connecting passageway at Narita Airport Terminal 3

4. Utilizing Colors to Showcase Cultural Characteristics

In addition to conveying different psychological meanings and guiding people's feelings, colors often become a cultural hallmark of a particular region. Influenced by factors such as history, culture, religion, and climate over many years, different countries and regions develop unique color combinations and symbolism. Because colors can reflect a region's cultural traits and carry symbolic meanings, airport terminal design can use typical colors to showcase a nation's or a region's cultural and character features.

Case Study: Athens International Airport in Greece
The blue civilization of the Aegean Sea

As an integral part of Greece's natural and human geography, Santorini Island represents the islands in the Aegean Sea. Its natural landscape comprises the azure Aegean Sea, brown volcanic rocks, and deep blue skies. The buildings, predominantly white, harmoniously blend with the natural environment. The extensive use of deep blue on roofs, window frames, and door signs allows the buildings to engage fully with the surroundings, creating a distinctive regional character. (Fig. 5-54)

The predominant color impression of Athens International Airport is "the omnipresent blue." The airport's signage system, public facilities, indoor advertisements, architectural components, and even door handles are adorned with a vibrant yet subtle cobalt blue. Blue is not used extensively but rather as accents in various areas. This application combines people's expectations of Greece, the Aegean Sea, blue rooftops, and ancient Greek civilization, providing passengers with a direct sensory experience of Greek culture. (Figs. 5-55, 5-56)

Figure 5-54 The blue civilization of Santorini Island

Figure 5-55 The blue facade of Athens International Airport terminal

Figure 5-56 Athens International Airport terminal, cobalt blue accents on a medium gray and white tone

Case Study: Guelmim Airport in Morocco

Light and shadows under perforated panels reflecting regional charm

The terminal facade of Guelmim Airport in Morocco is composed of a series of perforated panels in red and orange hues. This design allows external light to create star-pattern shadows indoors. When combined with sunlight, the colors emphasize the region's unique decorative theme (Fig. 5-57). The red and yellow facade complements the surrounding red earth and desert

Figure 5-57 The hollow pattern of walls of the Hassan II Mosque in Casablanca, Morocco

environment, reflecting the predominant colors of local Moroccan architecture. The perforated metal panels, casting starry light and shadows, resonate with the intricate patterns of stained glass windows in Moroccan mosques. This design beautifully represents the city's image and the cultural charm of local religious customs. (Fig. 5-58)

Figure 5-58 Guelmim Airport, the facade with distinct regional features harmonizing with the natural landscape and soil tones

5.2.3.2 Innovative Use of Materials

Architecture is often described as "solidified music," with materials acting as the notes that compose the melody. These materials possess essential characteristics such as materiality, naturalness, temporality, and regionality, forming the basis of architectural design. From the early use of stone, wood and brick, to modern materials like steel, concrete and glass, and to contemporary new materials, the continuous updates and innovative application of materials have always accompanied the evolution of building technology. In architectural design history, material use has skillfully combined culture and technology, imparting symbolic meanings and guiding the innovation of architectural design concepts and ideas.

As modern buildings, terminal buildings typically use steel for structural foundations, glass curtain walls for facades, and metal panels for solidity, reflecting a modern, simple, and efficient architectural expression. However, as the architectural industry progresses, designs that overly rely on these materials increasingly lack resonance with regional cultures.

Therefore, material innovation is crucial for the innovative design of terminal buildings. Different materials offer unique texture characteristics, providing distinct visual and tactile experiences for terminal users. Skillful use of materials can also echo regional cultural characteristics, creating a rich and nuanced viewing experience.

1. Innovation in Traditional Material Language and Application Methods

Traditional materials have a long history and mature technology, representing forms with distinct regional cultural characteristics that can easily inspire a sense of identity and belonging among passengers. Innovative terminal design can be achieved by looking at traditional materials from a new perspective, improving them with new technologies, and modernizing their use. Common traditional materials include wood, stone, and brick, often used in interior design to create a unique and delicate regional cultural ambiance. Modern materials such as steel, glass curtain walls, and metal panels convey simplicity and efficiency, while concrete materials are often used to present geometric shapes, enhancing the sense of ceremony. Exploring new ways to use existing materials' properties and combining different materials can stimulate innovative thinking and produce varied design effects.

Case Study: Nelson Airport Terminal in New Zealand
Locality in its large-span timber structure

Located in Wellington, the capital of New Zealand, Nelson International Airport handles 1.2 million person-times of passengers annually and covers an area of 5,300 square meters. Local timber is used for both the structure and interior decoration, integrating the terminal design with the natural landscape.

The combination of large-span timber structures and resilient seismic design has made Nelson Airport a standout in terminal architecture, serving as a model for sustainable airport terminal construction. The roof's well-designed timber elements provide an impressive visual contrast to the building's regular flat elevation. The roof surface units are arranged along the terminal's long axis, undulating in a regular pattern, echoing the surrounding natural mountains. (Fig. 5-59)

The interior ceiling material is a design highlight, with natural timber components creating a cozy atmosphere of harmonious coexistence between humans and nature. The interior's warm and elegant color scheme, along with the uniform texture of exposed wood structures, unifies the complex functional spaces. The natural properties of wood, such as texture, grain, and color, convey a straightforward and rustic message reflecting the design concept. This concept uses natural materials to echo the relationship between the terminal and the surrounding forested landscape, achieving a harmonious coexistence with the local natural environment. (Fig. 5-60)

Figure 5-59 Facade of Nelson Airport terminal

Figure 5-60 Interior space of Nelson Airport terminal

2. Practical Exploration on New Materials

The development of new materials provides more possibilities for innovative design in terminal buildings. Current innovations in building materials mainly involve synthetic, biomimetic, and smart building materials. However, at present, these new materials have not yet been widely used in the field of airport construction, offering significant potential for exploration and broad application prospects. This section introduces a terminal building case that utilizes the new materials UHPC (ultra-high-performance concrete), highlighting the advantages of new materials in practical applications and offering more possibilities for design innovation.

Case Study: Rabat-Salé Airport in Morocco
UHPC decorative components of the facade

The Rabat-Salé Airport features a single-level terminal building covering 5,200 square meters. The building's facade, measuring 1,600 square meters, consists of a traditional cultural pattern of the region — Moroccan Stars. This facade serves both as the building envelope and as a sunshade.

The facade is constructed using white concrete mesh panels spliced with organic fibers, each panel measuring 1.75 meters in width and 3.75 to 5.25 meters in height, with a thickness of 10 centimeters and a 70% perforation rate. This project marks the first use of UHPC in Africa. The material allows for the creation of a lightweight mesh structure, presenting an airport terminal image that reflects traditional Arab architectural characteristics. (Figs. 5-61, 5-62)

Figure 5-61 Mesh panel made of UHPC comprising the terminal facade

Figure 5-62　UHPC creates a lightweight skin of the building while maintaining structural strength

5.2.3.3　Innovative Use of Architectural Constructions

The construction system of a terminal building encompasses how various parts of the building are constructed and how these parts are combined. This design is influenced by factors such as the external environment, user demands, building technology, and project budget. It must also consider functional use, aesthetic value, technical efficiency, and other design aspects.

Meeting functional requirements is fundamental to terminal building construction design. Most terminals follow a highly efficient, single-manufacturing model at the construction aspect, prioritizing budget saving at the economic aspect. This prevailing design method often falls short in achieving structural design innovation under high standards of integration. To address this issue, designers can start from the user's perspective to find innovation points in constructive design. This approach includes fulfilling basic usage requirements and technical and economic constraints while emphasizing cultural connotation, visual performance enhancement, and the removal of redundant elements, thereby improving the comprehensiveness and advancement of constructive design.

Innovative design in detailed construction of terminal building can enhance users' experience. By researching the composition of each building part, architects can make buildings more approachable, deconstructing the huge volume of terminals to a human-friendly scale. Additionally, examining how these parts integrate can make the design language more coherent and enriched, reflecting local context and unique design intentions. Therefore, detailed construction innovation is significant for the design of small- and medium-sized airport terminals.

Case Study: Terminal 2 of Mactan-Cebu International Airport in the Philippines
Wood construction echoing local dwellings

To serve the second largest airport in the Philippines, the new terminal design of Mactan-Cebu International Airport fully considers the functionality of a transportation hub while deeply rooting itself in the local context and resonating with traditional Philippine architectural forms. The terminal resembles local dwellings with high-pitched roofs and low eaves, and its internal architectural construction also reflects regional characteristics. With an annual passenger throughput of 3.75 million person-times and covering 53,000 square meters, the terminal is a two-story configuration. Philippine traditional dwellings are easily constructed on the ground or over shallow water, supported by poles one to two meters high, with sloping lightweight roofs and an elevated space below. This design provides good air circulation and protection from floods, snakes, and insects. The terminal's roof construction mimics these local dwellings, supported by continuous 30-meter spans of laminated wood arches made from local spruce. The construction process, mastered by local craftsmen, ensures the low cost and high quality of future maintenance. This design adheres to the principles of simplicity, efficiency, and green sustainability in transportation architecture while imbuing the components with local cultural significance. (Figs. 5-63, 5-64)

Figure 5-63 Construction details echoing traditional dwellings within Terminal 2 of Mactan-Cebu International Airport

Figure 5-64 Wooden construction details of regional characteristics without losing sense of modernity

Case Study: Prince George Airport in Canada
Human-scaled construction in a small community airport

As a low-traffic community airport with a small terminal, Prince George Airport focuses on unique architectural construction details to highlight the advantages of small terminal designs. The boarding lounge features a well-lit glass atrium integrated with structural and shading details, allowing natural light to create rich shadow effects throughout the day. (Fig. 5-65)

The designers customized curtain wall connectors to make the wood-based curtain wall system lighter and the detailed language more streamlined and powerful. The unique point-fixed glass

Figure 5-65 Light-weight curtain wall and atrium for natural lighting of Prince George Airport terminal

Figure 5-66 The construction details: minimalist joints

system penetrates only the insulating unit's inner pane, preventing thermal bridging, enhancing aesthetics, and ensuring the building's sustainability. (Fig. 5-66)

5.2.3.4 Summary of Detailed Construction Innovation

Colors, materials, and construction details are fundamental components of innovative design in airport terminal. In terminal design, innovative detailing is crucial, not only reflecting the architect's creativity but also effectively highlighting the airport's uniqueness. Through meticulous consideration of human scale and refined material application, these detailing designs convey the airport's modernity, regional identity and future-oriented outlook.

5.2.4 Summary of Case Studies on Innovative Designs of Small- and Medium-Sized Airport Terminals

This chapter synthesizes innovative design methods for small- and medium-sized airport terminals from three aspects: external form, internal user experience, and detailed constructions (Fig. 5-67).

External form: Innovative external designs can derive from natural elements and humanistic attributes. "Innovation based on natural elements" involves drawing on topography, adapting the natural climate, and adopting bionic abstract form. "Innovation based on humanistic attributes" includes abstract

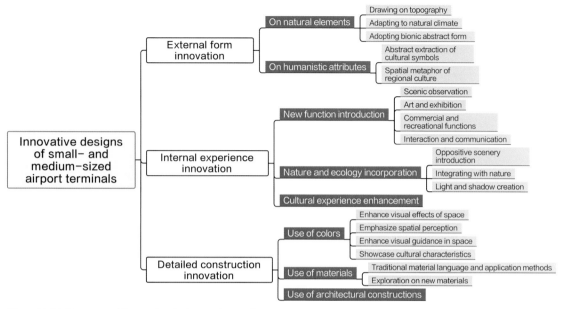

Figure 5-67 Summary of innovative design methods of small- and medium-sized airport terminals

extraction of cultural symbols, and spatial metaphor of regional cultures.

Internal experience: Innovation in internal space focuses on new function introduction, nature and ecology incorporation, and cultural experience enhancement. "New function introduction" encompasses features such as scenic observation, art and exhibition, commercial and recreational functions, and interaction and communication. "Nature and ecology incorporation" integrates oppositive scenery introduction, integration with nature, and light and shadow creation. "Cultural experience enhancement" adds cultural functions within the terminal's interior spaces.

Detailed construction: Innovative detailed construction designs emphasize color usage, material application, and intricate structural elements.

These three innovative design approaches are interwoven and mutually influential, advocating for a holistic design philosophy throughout terminal development. Integration across external form, interior experience, and detailed construction ensures a unified approach that enhances the terminal's functionality and aesthetic appeal.

TENDENCY

FOUR GENERATIONS OF REGIONAL AIRPORT
TERMINAL AND INNOVATIVE PRACTICES OF
THE FOURTH GENERATION
(ERA OF EXPERIENCE)

6.1 Four Generations of Regional Airport Terminal Design

In the more than seventy years since the founding of the People's Republic of China, the design and construction concepts of small- and medium-sized airport terminals in China have gone through several stages of development, each with its own distinctive features. The author has conducted a statistical study on the design of domestic small- and medium-sized regional airport terminals. By combining this with the design features of outstanding cases from both domestic and international sources, the development of design concepts of Chinese small- and medium-sized regional airport terminals can be roughly divided into four phases based on design concepts and focal points: the eras of function, style, culture, and experience.

The Era of Function coincided with the early stage of development of China's aviation industry, and the requirements for the design of terminal buildings at regional airports were not high. The design of terminal buildings was oriented towards functionality, focusing only on meeting the basic operational processes of the terminal. Terminal buildings in this period generally lacked the distinctive qualities of transportation architecture, and their design methods and stylistic language did not exhibit typical airport characteristics, often resulting in buildings that looked "unlike a terminal."

In the Era of Style, with the rapid growth of aviation demand for regional airports, the construction volume of domestic small- and medium-sized regional airports also showed explosive growth. The importance of airports as gateways and iconic landmarks of different cities was increasingly emphasized. The design focus of domestic regional airport terminals gradually shifted towards terminal aesthetics. On the basis of meeting the needs of functions and processes, terminal design aimed to create urban gateways, with increasing emphasis on image and aesthetics. The features of terminal buildings designed in this period include: emphasizing the building volume and scale, designing small- and medium-sized airport terminals with the design concept of large airport terminals; adopting large roof architectural shapes; exaggerated and unsubtle form language; lack of individuality in modeling and concepts.

In the Era of Culture, with the significant development of tourism in various small- and medium-sized cities in China and the emphasis of local governments on local culture, the expression of regional culture and characteristics has become a new demand in airport terminal construction. In this era, the design of regional airport terminal buildings has shifted from simply emphasizing aesthetics and image to highlighting regional culture and characteristics as urban cultural symbols. The features of terminal buildings of regional airports in this period mainly include: the excavation of regional cultural symbols as an important shaping concept; the architectural image reflecting certain regional differences; and the expression of cultural symbols primarily embodied in the architectural shape, with the collage of colors and cultural symbols being more common. There are many excellent cases of airport design in the cultural era, such as the terminal building of Shennongjia Airport in Hubei, but there are still many designs "for the sake of the region," blindly piling up or misusing regional cultural elements.

In the Era of Experience, with the rapid development of Internet technology, the sense of experience and communication has become a new concern for passengers in airport terminals. In this brand-new era, not limited to external modeling, how passengers can have different experiences and feel different spaces in airport terminals has become a new exploration direction for small- and medium-sized airport design. The characteristics of terminal buildings in this era mainly include: deep excavation of regional culture and clever extraction of cultural elements; emphasizing the display of cultural elements through the clever shaping and expression of modeling and space, rather than a mere collage of cultural symbols; integration of external modeling and internal space design; combining with the passenger process design to create a distinctive sense of regional cultural experience; and emphasizing the organic and close integration of the innovative experience with the passenger process.

In recent years, the author's team has responded to the new requirements in era of experience for airport terminal design by focusing on cultural experience. A series of innovative airport terminal design practices have been carried out. The following section will select some typical cases and analyze the specific design ideas and methods.

6.2 Innovative Design Practices of the Fourth-Generation Regional Airport Terminal

6.2.1 Xizang Dingri Airport Terminal: The "Snowy Eagle" Gazing at Mount Everest

6.2.1.1 Overview: The Airport Closest to Mount Qomolangma

Dingri Airport is located in the Xizang Autonomous Region, just 33 kilometers from Dingri County and only 52 kilometers from the Everest Base Camp, making it the closest airport to Mount Qomolangma in China. The airport site is at an elevation of 4,312 meters, ranking as the fourth highest airport in the world (Fig. 6-1). The airfield level is 4C, with a planning target year of 2030 to serve an annual passenger throughput of 250,000 person-times. It plans to have four Class C parking stands, and the terminal will be a one-and-a-half-story design, covering an area of approximately 8,000 square meters. In this planning, the airside of the terminal includes runways, taxiways, and apron, while the landside consists of the terminal area and working area. (Figs. 6-2 ~ 6-4)

Figure 6-1 Surrounding environment of Dingri Airport Site

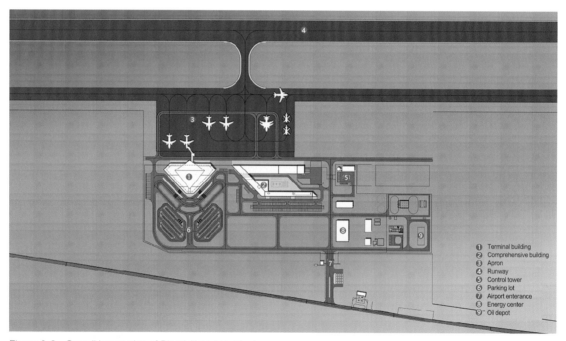

① Terminal building
② Comprehensive building
③ Apron
④ Runway
⑤ Control tower
⑥ Parking lot
⑦ Airport enterance
⑧ Energy center
⑨ Oil depot

Figure 6-2 Overall layout plan of Dingri Airport terminal area

Figure 6-3　Bird eye view of the landside of Dingri Airport terminal area

Figure 6-4　Bird eye view of the airside of Dingri Airport terminal area

6.2.1.2　Soul of Design: Drawing Inspiration from Nature

In architectural design, it is halfway to success when the "soul" of the building is grasped. At the beginning of the design of Dingri Airport terminal, the design team focused on how to present a unique and unforgettable building for passengers in such a magnificent, pure and sacred environment, and how to integrate the efficient and simple qualities of transport architecture with the rich cultural heritage of Xizang. More importantly, how to capture the "soul" of the Dingri Airport terminal design? Is it a reproduction of traditional Zang architecture such as the Potala Palace (Fig. 6-5), or is it lays in a different approach?

The design team believes that "imitation can never surpass." The architecture can pay homage to classic Zang-style elements through color and details, and it must primarily reflect simplicity, elegance, and efficiency while also integrating the essence of local culture. Thus, the design of the Dingri Airport terminal does not follow traditional Zang architectural styles; instead, it learns from nature and seeks inspiration from the unique natural environment of Xizang region. Ultimately, the design team chose the concept of the "Eagle of Snowland," a symbol of bravery, strength, and fortitude, using modern and minimalist design techniques to embody a robust and soaring architectural spirit. (Figs. 6-6, 6-7)

Figure 6-5　Potala Palace

Figure 6-6　Bird eye view of a certain airport terminal in Xizang

6.2.1.3　Terminal Planar Configuration: Triangle

The architectural shape relies on the support of its configuration. The original layout of the terminal was a conventional shape of rectangle, which was difficult to align with the image of "Eagle of Snowland." After studying many Chinese and foregin regional airport terminals, the design team found that the vast majority of small- and to medium-sized terminals featured rectangular shapes or their variations, with very few adopting circular forms, which contradicts the eagle's form. After considering the internal workflow and the concept of the eagle, a "triangle" configuration was boldly chosen for Dingri Airport. This choice perfectly achieved an unity of "soul-configuration-shape-function" of the terminal building. (Fig. 6-8)

Figure 6-7　Eagle of Snowland

Figure 6-8　Schematic diagram of the triangular configuration

6.2.1.4 Landside Transportation System: Simple and Smooth

The landside transportation system at Dingri Airport is organized separately for the terminal, the comprehensive area, and work area, with clear and distinct vehicle flow patterns (Fig. 6-9). In front of the terminal, there are two layers of curbsides: the inner curbside is for passenger drop-off, while the outer one is for by-pass traffic. The outer side of the curbside borders the parking lot. The two right-angled edges of the triangular terminal are arranged as follows: the upstream curbside accommodates private vehicles, taxis, and online taxis for passenger drop-off, while the downstream curbside is designated for airport shuttles to drop off and pick up passengers. The parking lot is organized for private vehicles to pick up passengers. Given the low annual passenger throughput, the length of the drop-off curbside in front of the terminal is more than sufficient to accommodate passenger flows during peak hour based on the capacity of a C-class aircraft. (Fig. 6-10)

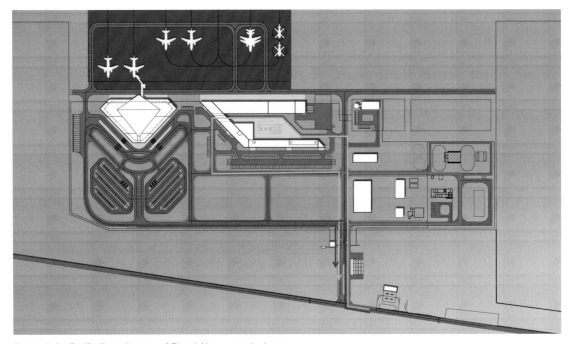

Figure 6-9   Traffic flow diagram of Dingri Airport terminal area

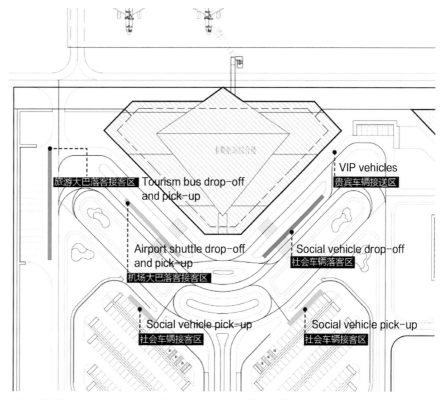

Figure 6-10   Landside transportation organization of Dingri Airport terminal area

6.2.1.5   Functional Flows: Clear at a Glance

      Due to the triangular configuration, the interior of the terminal naturally forms an L-shaped space for arrival and departure halls. The check-in area is on the right while the arrival area is on the left, and the security checkpoint is in the middle. The baggage claim hall and baggage handling rooms are located on the left and right sides, respectively. The boarding lounge is situated in the central area behind the security checkpoint, while the second floor features the boarding corridor, the preserved area for near parking stands in future, and office space. The VIP lounge is located on the first and second floors of the right side of the terminal. (Figs. 6-11 ~ 6-14)

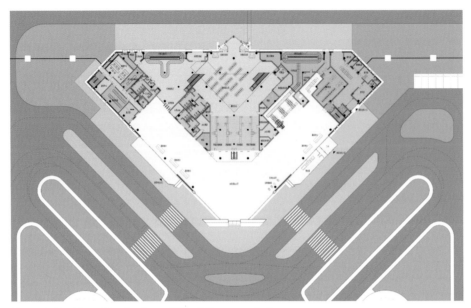

Figure 6-11   Plan of the first floor of Dingri Airport terminal

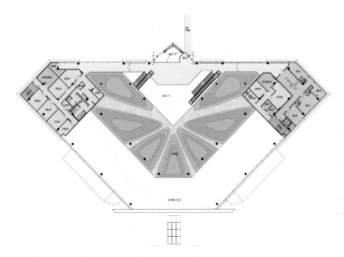

Figure 6-12   Plan of the second floor of Dingri Airport terminal

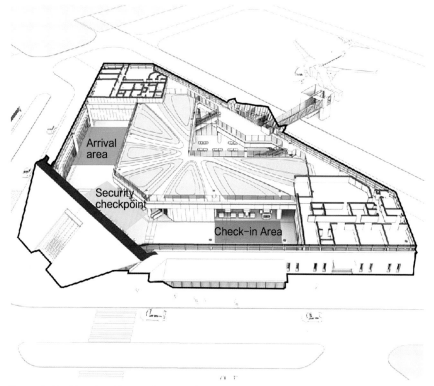

Figure 6-13　Functional layout diagram of Dingri Airport terminal

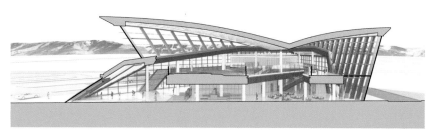

Figure 6-14　Sectional diagram of Dingri Airport terminal

6.2.1.6　Architectural Form: Eagle of Snowland

The shape of the terminal resembles an eagle poised to take flight, resting on the snowy plateau and echoing the sacred mountain, expressing a reverence for nature and a yearning for Mount Qomolangma. The overall architectural form is symmetrically arranged along a central axis, with the forward-tilting large roof symbolizing the proud head of the eagle, while the secondary roofs on either side resemble the eagle's wings, ready to soar. The architectural language is concise and refined, exuding a strong modern feel while capturing and expressing the image of the "Eagle of Snowland" with grandeur, sophistication, and dynamism.

The terminal is primarily constructed with silver metal panels and glass, creating a cohesive facade, while red aluminum panel accents and local stone walls at the entrances pay homage to traditional Zang-style architecture. (Figs. 6-15 ~ 6-18)

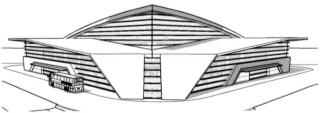

Figure 6-15　Conceptual form of the "Eagle of Snowland" for Dingri Airport terminal

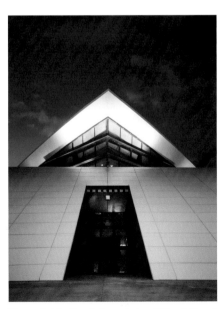

Figure 6-16　Front view photo of Dingri Airport terminal

Figure 6-17 Close shot of the front of Dingri Airport terminal

Figure 6-18 Side view of Dingri Airport terminal

6.2.1.7　Interior Space: Eagle Feathers and Khata

　　The terminal's design seamlessly integrates its external form and internal space. The large ceiling in the boarding lounge serves as the most dominant visual element, extending the creative inspiration of the "Eagle of Snowland": The ceiling features triangular panels that are slightly tilted and beveled, combined with varying light effects, creating an impression of "eagle feathers." This reinforces the design concept of eagle within the interior, achieving a clean, refined look that resonates with the essence of modern transportation architecture, and creates a spatial flow directing attention toward the sacred mountains, providing clear visual guidance for passengers while enhancing their experience. Moreover, the triangular texture aligns with the building's form, facilitating standardized and modular construction that adds a rhythmic beauty to the space.

　　Another important area is the arrival and departure halls, which must adequately express the regional culture. As a space for welcoming and sending off passengers, it symbolizes hospitality and blessings. The design team created an L-shaped space with a continuous ceiling design, featuring layered white metal strips integrated with linear lighting. This symbolizes the terminal presenting a white Khada to passengers, embodying a warm greeting.

　　Cultural experiences are reflected not only in forms and spaces but also in the pure and modern detailing of the architecture. For detail construction design, the design team incorporated elements and colors from the Zang culture, achieving harmonious fusion of modern transportation architecture and local culture. For instance, the design cleverly extracted details and colors from Zang-style palace architecture and used them in the first and second floors, combining red eaves, yellow beams, and decorative lines to convey a classical yet modern architectural quality.

Figure 6-19　Imagery of Khada and eagle feathers in the Zang culture

The conceptual intent and design of the airport profoundly interpreted local culture, delivering an excellent experience for visitors. The terminal engaged in a harmonious dialogue with its surroundings, making the architecture an integral part of the stunning landscape. (Figs. 6-19 ~ 6-24)

Figure 6-20 Interior rendering of the second floor of Dingri Airport terminal

Figure 6-21　Interior rendering of the second floor of Dingri Airport terminal

Figure 6-22　Interior rendering of the second floor of Dingri Airport terminal

Figure 6-23　Interior rendering of the boarding lounge on the first floor of Dingri Airport terminal

Figure 6-24　Interior rendering of the departure and arrival halls on the first floor of Dingri Airport terminal

6.2.1.8　Signage and Identity: Prayer Wheels and Zang-style Dressing

Signage is a crucial element for guiding processes in transportation architecture. In addition to incorporating international standards for signs including forms, fonts, and colors, the design team added delightful surprises for passengers to discover and enjoy: For example, the tops of the signage poles are inspired by prayer wheels, symbolizing blessings; the gender signs for restrooms use simplified, abstract representations of people in traditional Zang-style dressings, reflecting the spirit and cultural essence of local people. (Fig. 6-25)

Figure 6-25 Abstract representation of prayer wheels and Zang-style dressing elements in signage system

6.2.1.9 Outdoor Landscape: Mani Stones and Climber Sculptures

The design of the outdoor landscape really challenged the design team. At an altitude of over 4,300 meters in the snowy plateau, various ornamental plants were unsuitable. The design team had to abandon the idea of using greenery for landscaping and opted for hard materials and localized water features instead. Drawing inspiration from analyzing passenger demographics provided a sudden solution to the landscape issue — "Sculptures of Mount Qomolangma Climbers." This not only perfectly fits the site but also offers more creative opportunities. (Figs. 6-26 ~ 6-28)

Additionally, in the entrance area of the airport, the design placed a Mani stone pile at an appropriate location along the path to the terminal, indicating key information such as altitude and distance to the Everest Base Camp, which serves as an appropriate landscape presentation. (Fig. 6-29)

Figure 6-26　Rendering of the climber sculptures in front of Dingri Airport terminal

Figure 6-27　Photo of the climber sculptures

Figure 6-28　Side view of Dingri Airport terminal

Figure 6-29　Rendering of the Mani stone pile in Dingri Airport terminal area

6.2.1.10 Unique Experience for Passengers: "Climbing the Mount, Showcasing Passion"

As the airport closest to the Everest Base Camp, most passengers flying to Dingri Airport have a singular purpose: to climb Mount Qomolangma and showcase their passion. Therefore, the design must respond to the travelers' initial impressions with the most direct spatial experience. The soaring, dynamic "eagle head" structure allows travelers to admire the sacred snow-capped mountains through elevated side windows as soon as they enter the terminal, enhancing and elevating the climbers' lofty emotions.

This project is both a crystallization of the modern expression of local Zang culture and a comprehensive innovation exploring fundamental issues such as the relationship between airport space and passenger experience.

6.2.2 Lancang Jingmai Airport Terminal 2: "Scenic Green Valley, Prosperity Ahead"

6.2.2.1 Overview: Four Seasons like Springtime

Lancang Jingmai Airport is located in the southwestern part of Lancang County, Pu'er City, Yunnan Province, at an elevation of 1,350 meters. It is approximately 127 kilometers from Pu'er City in a straight line and is situated at the center of the "Pu'er Green Triangle" tourism area, which includes Lancang, Ximeng, and Menglian. Pu'er City boasts abundant forest resources, with a forest coverage rate of 64.9%, earning it the nickname "Pearl of the Green Sea." The average annual temperature is 18.2℃; the coldest month averages 12.9℃, while the hottest month averages 22.1℃, making this region enjoy a favorable climate through the year.

Before the expansion, Lancang Jingmai Airport had only one T1 terminal with an annual passenger throughput of 250,000 person-times. With 2045 as the target year, this design will expand the airport by adding five near parking stands, three remote parking stands, and expanding the Terminal 2 to the east of the Terminal 1, and the annual passenger throughput of the new terminal is estimated to be 1.35 million person-times. That is, after the completion of the Terminal 2, the airport's annual passenger throughput will reach 1.6 million person-times with the airfield level being 4C. Terminal 2 echoed Terminal 1 using rectangular plan configuration, with a width of about 180 meters, depth of about 46 meters, the highest roof of about 20 meters; its sectional configuration is the one-and-a-half-story type, with the building area of about 12,000 square meters. (Figs. 6-30 ~ 6-32)

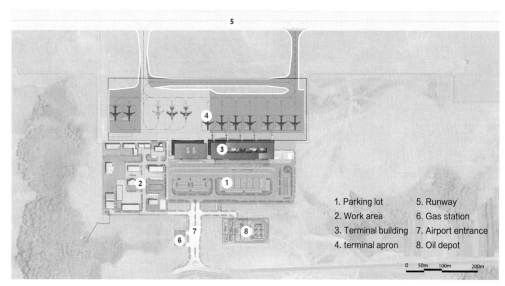

1. Parking lot
2. Work area
3. Terminal building
4. terminal apron
5. Runway
6. Gas station
7. Airport entrance
8. Oil depot

Figure 6-30 Overall layout plan of Lancang Jingmai Airport

Figure 6-31 Bird eye view of the landside of Terminals 1 and 2 at Lancang Jingmai Airport

Figure 6-32　Bird eye view of the airside of Terminals 1 and 2 at Lancang Jingmai Airport

6.2.2.2　Five Key Features of Terminal Design

1. Reasonable Zoning and Flexible Expansion

The terminal is divided into two main functional areas: On the first floor, the terminal is divided into the departure zone and arrival zone through a central landscaped courtyard. The departure zone is located on the east side of the courtyard and includes the departure hall, check-in area, and security checkpoint. After passing security check, passengers can take escalators to the second-floor near boarding lounge on the east side, with clear guidance provided. The remote boarding lounge is arranged on the west side of the security check passage. The arrival zone is located on the west side of the courtyard and mainly consists of the baggage claim hall and arrival hall. The government and business VIP lounge is situated on the east side of the terminal, with a separate entrance and exit that can be connected or separated from the departure hall. The second floor features the near boarding lounge, arrival corridor, and outdoor airside landscape courtyard. (Fig. 6-33)

To accommodate future passenger growth, the terminal's facilities must have flexible expansion space. Based on the current facilities in the terminal, space for expanding check-in, security check, and other facilities is reserved to meet the service demand for a future capacity of 1.8 million person-times passengers annually.

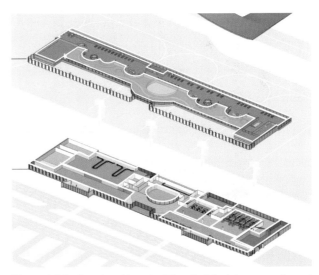

Figure 6-33　Functional layout of Terminal 2 at Lancang Jingmai Airport

2. Innovative Experience: The Jingmai Green Valley and the Gourd in the Garden

A climate akin to spring all year round and the famous tea gardens are the very first impression for travelers arriving in Lancang. The design team has creatively developed a 145-meter-long airside green valley on the second floor of Terminal 2. This cleverly incorporates Lancang's green landscape into the terminal space on a small scale; at the same time, the gourd, namely the cultural totem of the Lahu People, is artistically integrated into the space, extending from the boarding lounge into the ecological green valley. It serves as a lively recreational area in the boarding lounge, reflecting the gourd's auspicious and peaceful symbolism in Lahu culture.

The boarding lounge and the green valley are fully integrated via the gourd-like spaces, each complementing one another. The outdoor space of the ecological green valley is also part of the airside waiting area, allowing for a seamless transition to the gourd space, through which travelers do not have to stay in the enclosed waiting space, but can step into the green valley to fully enjoy the spring-like natural environment. (Figs. 6-34 ~ 6-37)

Figure 6-34　"Green Valley" concept for Terminal 2 at Lancang Jingmai Airport

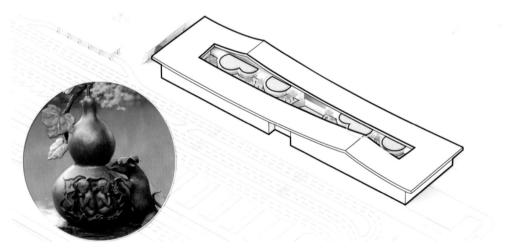

Figure 6-35　Abstraction of the "gourd" element and spatial transformation in Terminal 2 at Lancang Jingmai Airport

Figure 6-36　Perspective renderings of the landscape courtyard in Terminal 2 at Lancang Jingmai Airport

Figure 6-37　Perspective rendering of the landscape courtyard in Terminal 2 at Lancang Jingmai Airport

3. Formed by the mountain, and highlighted with culture

As the "aerial gateway" of a city, the airport terminal also serves as a shining city business card. Its facade is the first impression travelers have when entering the terminal. Surrounded by mountains, the design of the terminal's roof reflects the shape of the landscape, blending with nature. The elegant, sweeping curves of the roof slightly lift on the east side of the landside facade, indicating the location of the departure hall. On the airside, the ridge rises on the west side of the facade, suggesting the welcoming space of the terminal. The north-south shifting of the ridgeline adds sense of flowing to the building, while the eaves and south facade are presented with a zigzag glass curtain wall, subtly evoking the patterns of the local Lahu clothing. The architectural language expresses regional cultural symbols in a restrained yet elegant manner. (Figs. 6-38, 6-39)

The expansion project must directly address the relationship between the old and new buildings. Since the width of Terminal 2 is twice that of Terminal 1, it is essential to avoid a drastic scale difference

Figure 6-38   Mountain form for roof shaping of Terminal 2 at Lancang Jingmai Airport

Figure 6-39   Cultural symbol extraction and application in the detail of Terminal 2 at Lancang Jingmai Airport

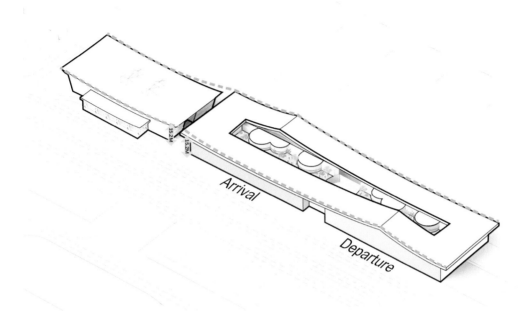

Figure 6-40　Overall relationship between Terminals 1 and 2 at Lancang Jingmai Airport

Figure 6-41　Perspective rendering of the main facade relationship between Terminals 1 and 2 at Lancang Jingmai Airport

Figure 6-42　Relationship between the roofs of Terminals 1 and 2 at Lancang Jingmai Airport from the airside

Figure 6-43　Relationship between the roofs of Terminals 1 and 2 at Lancang Jingmai Airport from the landside

Figure 6-44　Bird eye view of the overall relationship between Terminals 1 and 2 at Lancang Jingmai Airport

between them. The design employs three approaches: ① Terminal 2 features a curved roof that harmonizes with Terminal 1; ② The eaves of Tterminal 2 are aligned at the same height as T1; ③ The facade of Terminal 2 is divided into departure and arrival sections to reduce the visual scale. The newly constructed Terminal 2 sits humbly adjacent to Terminal 1, achieving a harmonious unity. (Figs. 6-40 ~ 6-44)

4. Commercial Planning: Industrial Value Enhancement

Commerce, as an important part of non-aviation revenue, has formed an effective organizational connection in this design, with the centralized commercial area mainly located around the atrium and the airside landscape courtyard. The combination of local culture promotion and display interfaces on passenger flow lines explores the value of local cultural industries, serving as an effective way to showcase tea culture, symbols of ethnic minorities, and handicrafts.

The leisure dining area combined with seating in the waiting area, retail shops, and commercial spaces both inside and outside the landscape courtyard form an interesting "point – line – surface" commercial space, greatly increasing the diversity of potential commercial types and also enhancing the connections between various businesses. (Figs. 6-45 ~ 6-48)

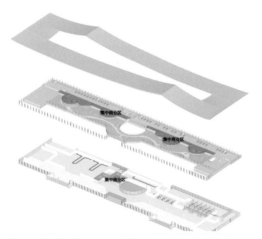

Figure 6-45 Commercial layout for Terminal 2 at Lancang Jingmai Airport

Figure 6-46 Interior rendering of the commercial area in Terminal 2 at Lancang Jingmai Airport

Figure 6-47 Interior rendering of the waiting corridor in Terminal 2 at Lancang Jingmai Airport

Figure 6-48 Perspective rendering of the commercial area beside landscape courtyard of Terminal 2 at Lancang Jingmai Airport

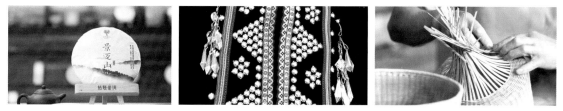

Figure 6-49 Intangible cultural heritage presented in Terminal 2 at Lancang Jingmai Airport

Retail business in the terminal can develop a product series of characteristic handicrafts and local gourmet food (Fig. 6-49), while specialty restaurants are mainly set in conjunction with the landscape courtyard and the needs of short-stay passengers.

5. Sustainable Design: Low-carbon, Energy-saving and Ecological Introduction

The terminal's curtain wall adopts Low-E glass with silk screen printing, forming a self-shading system, which ensures the transparency of the building while effectively reducing solar radiation and glare,

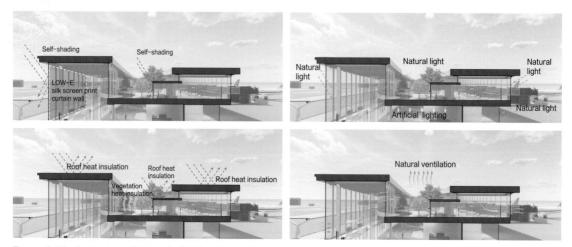

Figure 6-50 Natural ventilation, daylighting, thermal insulation, and passive shading of Terminal 2 at Lancang Jingmai Airport

thus lowering energy consumption. The construction is optimized to reduce thermal bridges, enhancing the insulation and thermal performance of the roof and exterior walls.

The indoor light environment is optimized and enhanced, taking into account the needs for shading and views, combined with landscape courtyard, and optimizing the design of facade windows and high side windows, ensuring ample and even natural lighting indoors. At the same time, lighting control is integrated with natural lighting to effectively reduce lighting energy consumption. The facade opening design is optimized based on wind pressure analysis to effectively organize natural ventilation, effectively reducing energy consumption in the transitional seasons. (Fig. 6-50)

6.2.2.3 Major Principles of Landside Transportation Organization: Public Transport Priority, Multiple Curbsides, and One-way Circulation

At Lancang Airport, the curbsides in front of the terminal building are divided into inner and outer layers. The inner curbsides serve the airport shuttle buses for passenger pick-up and drop-off and taxi pick-up, while also accommodating special vehicles; the outer curbsides are for taxis and social vehicle passenger drop-off. The parking lot has an inner curbside close to the terminal building (Fig. 6-51). Different types of vehicles are zoned and arranged on the different sections of curbside for unified

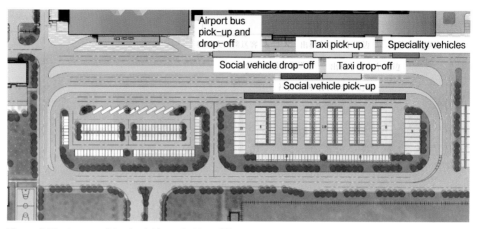

Figure 6-51　Layout of the landside curbsides of Terminal 2 at Lancang Jingmai Airport

management to improve operational efficiency. All sizes of social vehicles and online taxis are organized to pick up passengers within the parking lot.

Besides, this design adopts "one-way circulation" in landside transportation organization to avoid interweaving traffic and improve operational efficiency. To be specific:

Airport buses drop off passengers at a fixed area at the inner curbside and pick up passengers here to leave; taxis drop off passengers at the outer curbside, after which they proceed to the waiting area, and move to the inner curbside in front of the building to pick up passengers and leave; social vehicles drop off passengers at the outer curbside in front of the building or in the parking lot, and pick up passengers inside the parking lot then leave; as for VIP vehicles, departing VIPs are taken to the VIP lounge via a dedicated lane, and arriving VIP vehicles join the one-way circulation system in front of the building to leave. (Figs. 6-52, 6-53)

Figure 6-52　The one-way circulation landside transportation system of Terminals 1 and 2 at Lancang Jingmai Airport

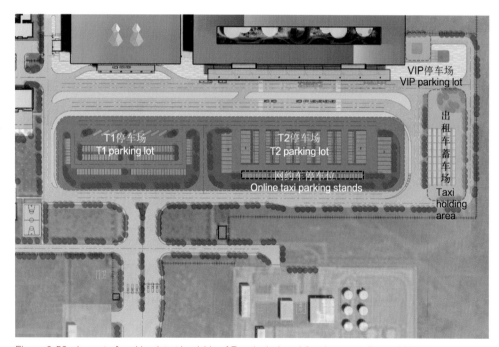

Figure 6-53　Layout of parking lot at landside of Terminals 1 and 2 at Lancang Jingmai Airport

6.2.2.4 Unique Experience for Passengers: "Entering Jingmai Green Valley to Have Fortune (Gourd)"

Lancang Jingmai Airport represents the author team's first attempt at designing a fourth-generation regional airport terminal building. The "unique" nature of the passenger experience, reflecting the airport's own characteristics, is what the team is pursuing. The design team focused on two key points: Firstly, abandoning the plain expression of "gourd pillars" and a "golden gourd" at the roof eaves of facade in Terminal 1; instead, they adopted a spatialized "gourd space" interpreted architecturally from the cultural symbol of Lahu People, allowing passengers to savor the beautiful Lahu legend that people come from gourds (Fig. 6-54). Secondly, it has innovatively designed an "airside ecological green valley." This courtyard is also part of the airside waiting area, seamlessly transitioning with the gourd spaces, allowing passengers to fully enjoy the spring-like natural environment all year round.

Upon the design of Terminal 2 of Lancang Jingmai Airport, Jingmai City is able to present its welcoming image to wide passengers with a unique impression. (Fig. 6-55)

Figure 6-54　Comparison of "gourd" imagery used in Terminals 1 and 2 at Lancang Jingmai Airport

Figure 6-55 Perspective rendering of the main facade of Terminal 2 at Lancang Jingmai Airport

6.2.3 Cangyuan Washan Airport Terminal[①]: Airport Museum in "Cave Paradise in Wooden Drum"

6.2.3.1 Overview: The Aerial Gateway to the Mysterious Wa Village

Washan Airport is located in Cangyuan County, southwestern Lincang City, Yunnan Province, approximately 20 kilometers from center of the county. It is situated in the core area of Lincang's initiative to create a "Mysterious Wa Village of the World." Cangyuan, known as the "Great Wa Mountain Area," is the largest settlement of the Wa People in China, as well as the birthplace, gathering place, and inheritor of Wa culture. It is also renowned as a world-famous community for the Wa People, a site of cliff paintings,

---

① This case is the proposal for bidding.

and a land of song and dance on the border.

Since the Terminal 1 commenced operations in December 2016, the passenger throughput of Cangyuan Washan Airport has rapidly increased, reaching 337,000 person-times in 2019, surpassing the designed passenger throughput of 270,000 person-times for 2020. With the annual growth in passenger volume and flight frequency, the current terminal has been operating beyond its capacity. Furthermore, to respond to national initiatives, improve the integrated transportation system, and promote local economy and tourism, Cangyuan Washan Airport started the expansion and renovation of Terminal 2 and the associated facilities.

This design aims for the year 2030, planning for an annual passenger throughput of 1 million person-times. It will add 4 near parking stands and 1 general parking stand, making the total of parking stands to 11 on the apron. A new 11,000-square meter Terminal 2 will be constructed, along with expanded transportation facilities in front of the terminal. Additionally, there will be expansions of office space, overnight accommodations for passengers, a cargo station, staff living areas, ground services, an emergency center, and power and fuel facilities on the reserved land. After the completion and operation of Terminal 2, Terminal 1 may consider serving international routes and VIP passengers in the future. (Figs. 6-56 ~ 6-59)

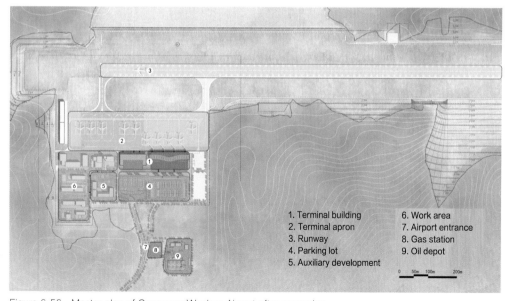

Figure 6-56   Master plan of Cangyuan Washan Airport after expansion

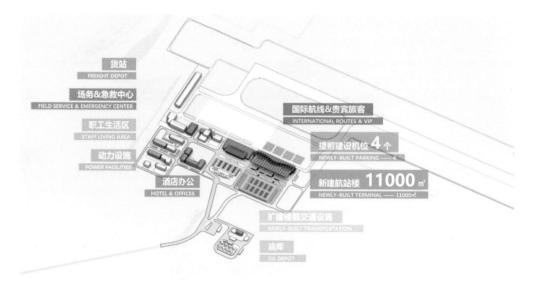

Figure 6-57　Functional layout of Cangyuan Washan Airport after expansion

Figure 6-58　Bird eye view of Terminals 1 and 2 from the landside of Cangyuan Washan Airport

Figure 6-59 Bird eye view of Terminals 1 and 2 from the airside of Cangyuan Washan Airport

6.2.3.2 Design Concepts: Cave Paradise in Wooden Drum, Airport Museum

1. Form out of Culture: Cave Paradise in Wooden Drum

As an unpolluted hidden gem, the beautiful and magnificent natural scenery of Cangyuan, along with its mysterious and ancient cultural charm, awaits every traveler to explore and experience. As the engine driving tourism development in Cangyuan and Lincang, Washan Airport is the first stop for guests arriving in Cangyuan, bearing the responsibility of introducing this mystical place to the world.

The vibrant and mysterious Wa culture is the most immediate impression for visitors to Cangyuan. Discovering and interpreting local culture is a key focus of this terminal space design. The design team sought cultural forms that are both representative and capable to cleverly integrate with architectural form and space. Thus, the "wooden drums" played by the Wa people during major festivals, along with the beautiful legend of the "Sigangli" (humans emerging from caves), representing the region's rich ethnic culture, became the two major inspirations of design. (Fig. 6-60) The "wooden drum" can be shaped, and the cave can be molded into an internal space. Ultimately, the design team decided "Wooden Drum for Form, Cave Paradise into Experience" as the theme of the terminal's space.

The second-floor waiting corridor features a striking "suspended wooden drum," creating a unique representation of the Wa culture in Cangyuan, resonating the secret sounds of Wa People to the world. (Fig. 6-61) The interior of the waiting corridor draws inspiration from the mythology of "Sigangli," using cave-like rock formations as a motif. Through architectural transformation of space and components, the cave exploring experience is expressed in modern architectural manner. In this 156-meter-long flowing

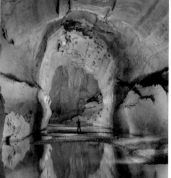

Figure 6-60　The "wooden drum" and "Sigangli" imageries

Figure 6-61 The "wooden drum" concept extraction and architectural transformation into waiting corridor space

Figure 6-62 The "cave space" concept extraction and architectural transformation into boarding lounge interior

Figure 6-63　Interior rendering of interior lighting in the waiting corridor of Terminal 2 at Cangyuan Washan Airport

space, travelers are provided with a unique experience during waiting (Fig. 6-62). In terms of lighting, the design boldly combined skylight with artificial lighting, which is not typically found in conventional boarding lounge. This ensures sufficient brightness indoors while maximizing the cave-like spatial experience intended by the concept. (Fig. 6-63)

2. Wa Culture Museum, Immersive Experience

Beyond exploring cultural symbols in terminal design and spatial layout, creating a unique cultural experience is essential for leaving a lasting impression on visitors.

The "Great Wa Mountainous Area" is rich in cultural tourism resources, with over 3,500 years of history in the Wengding primitive tribe, Cangyuan rock paintings, Guangyun Temple, and the lively "Moh-Ni-Hei" carnival, all reflecting deep historical and cultural legacies. The area also boasts stunning natural attractions such as the thousand-meter-long landscape painting corridor, Sigangli, sinkholes, and the Nangun River. However, these attractions are widely dispersed, making it challenging for visitors to explore them all in a short time. (Fig. 6-64)

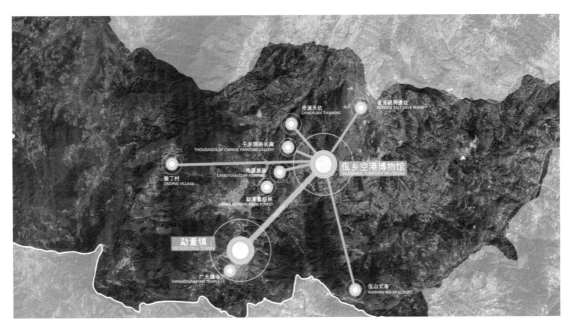

Figure 6-64 Distribution of attractions in the Great Wa Mountain Area

To address this, the design integrated the concept of a "museum" into the airport design, creatively proposing the idea of a "Wa Culture Airport Museum." (Fig. 6-65) The design team combined boarding lounge with exhibition spaces, cleverly incorporating the "Wa Culture Airport Museum" into the corners of the terminal's second-floor waiting corridor and the main building. Through cultural exhibits, interactive experiences, and multimedia presentations, they vividly showcase the rich history, natural landscapes, and ethnic customs of the enchanting Wa homeland.

Within this space, the concept of "Cave Paradise in Wooden Drum" is utilized to display Cangyuan rock paintings, Wengding Village, Sigangli, the wooden drum, the "Hair-swinging" Dance, and the "Moh-Ni-Hei" celebration. The exhibition halls blend seamlessly with the waiting areas, so travelers can easily step into the museum while waiting for boarding. To create a sense of mystery inside the airport museum, concrete wall materials are used, with a gap left between the wall and ceiling that allows light to shine on the rock paintings, producing a more mystical visual effect. (Fig. 6-66)

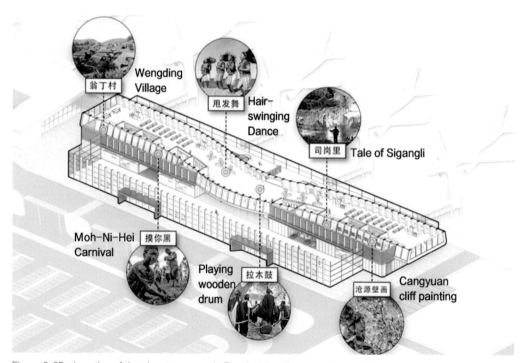

Figure 6-65 Location of the airport museum in Terminal 2 at Cangyuan Washan Airport

Figure 6-66 Interior renderings of the airport museum in Terminal 2 at Cangyuan Washan Airport

3. Color Design

After establishing the cultural experience concept of the "Airport Museum" and the spatial experience concept of the "Wood Drum Paradise," the design focused on enhancing the experience in detail. It revolves around the concept of the suspended wood drum, boldly employing colors such as wood yellow, dark red, and black, which not only emphasize the imagery of the wooden drum but also incorporate strong characteristics of the Wa People. Inside the waiting corridor (wooden drum), wooden structural ribs are used to create a unique sense of space reminiscent of the paradise in a wood drum, adhering to a coherent design logic of "concept – form – space – color." (Figs. 6-67 ~ 6-69)

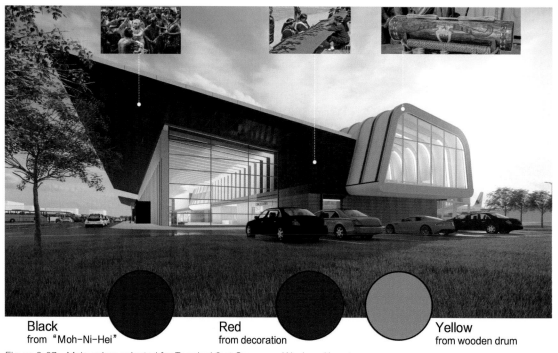

Figure 6-67　Main colors selected for Terminal 2 at Cangyuan Washan Airport

Figure 6-68 Color effect of the departure hall in Terminal 2 at Cangyuan Washan Airport

Figure 6-69 Color effect of the departure hall in Terminal 2 at Cangyuan Washan Airport

4. Relationship between Terminals 1 and 2

As an expansion project, the design also focused on the coordination and unity with the existing Terminal 1. The roof of Terminal 2 continued the architectural language of Terminal 1, creating a grand and unified airport image with the same roof slope. The facade treatment maintained a genetic connection with Terminal 2, being harmonious yet distinct, achieving an effect that is both modern and coordinated, and filled with a unique regional identity. (Figs. 6-70, 6-71)

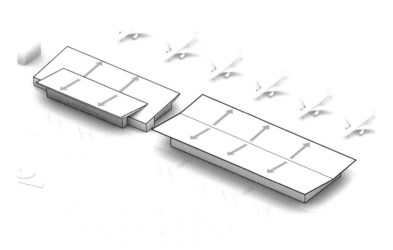

Figure 6-70 Schematic diagram of the relationship between Terminals 1 and 2 at Cangyuan Washan Airport

Figure 6-71 Perspective rendering of the front facade relationship between Terminals 1 and 2 at Cangyuan Washan Airport

6.2.3.3 Functional Flow and Standardized Construction: Rational Zoning, Cost Control

The new terminal adopts one-and-a-half-story structure with two main functional layers. The first floor features a central check-in area, with the departure zone located to the east of the check-in area, including the departure hall and security checkpoint. After passing through security check, passengers can take an escalator to the second-floor boarding lounge, with clear guidance provided. The remote boarding lounge is situated to the east of the security check passage. The arrival zone is located on the west side

Figure 6-72 Exploded axonometric diagram of Terminal 2 at Cangyuan Washan Airport

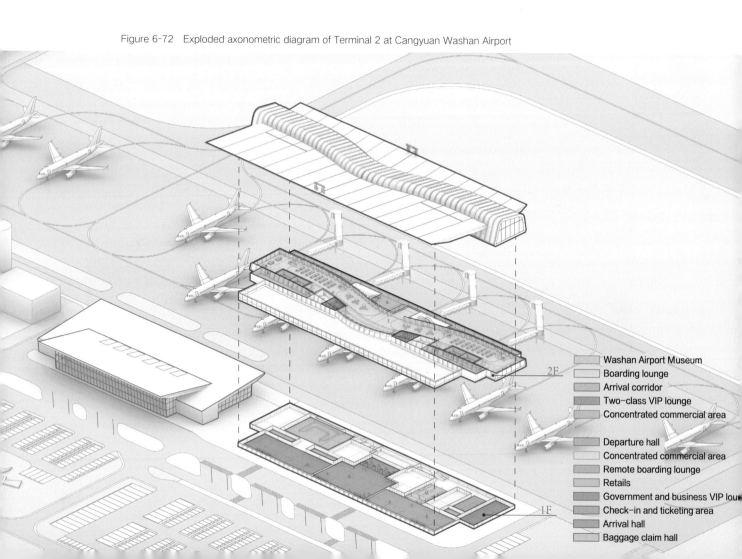

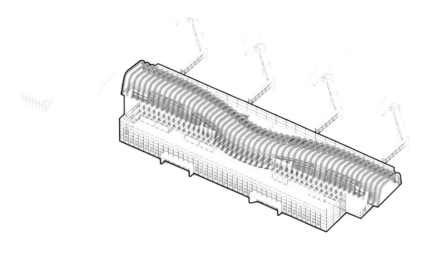

Figure 6-73 Modular standardization for architectural components of Terminal 2 at Cangyuan Washan Airport

of the check-in area, primarily comprising the baggage claim hall and arrival hall. The government and business VIP lounge is located on the east side of the terminal, with an independent entrance and exit that can be either separated from or integrated with the departure hall. The second floor includes the near boarding lounge, arrival corridor, and Wa Culture Airport Museum. (Fig. 6-72) Escalators for both departures and arrivals are centrally located on the second floor to minimize passenger walking distances.

  The regular main building and the curved waiting space use a unified modular system to standardize components, thus reducing costs and ensuring controllability, while achieving optimal spatial effects. (Fig. 6-73)

6.2.3.4 Commercial Planning: Industrial Improvement Helps Poverty Alleviation

The commercial planning of Terminal 2 is an important part of the design, and the commercial area provides an excellent window for local speciality products and handicrafts to go out of Washan. The design combines the airside and landside with passenger flow lines to form an effective commercial space system: the centralized commercial area is closely integrated with the waiting space and the Wa Culture Airport Museum; the local special products and handicrafts exhibition and cultural display spaces are interspersed in the waiting space to make the terminal building become an important window for the Cangyuan poverty alleviation achievements and the display of the intangible cultural heritage; the retail and catering space in the building is mainly combined with the passenger demand and arranged in the passenger departure and arrival line, which is convenient for consumption; the leisure commercial area is combined with the seating area to form a flexible and interesting commercial space, which greatly increases the diversity of commercial business. (Fig. 6-74)

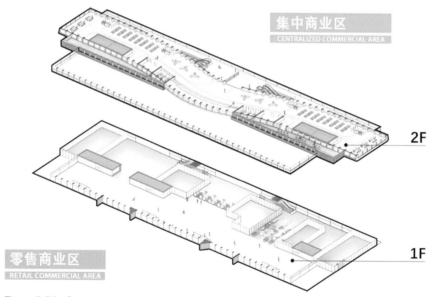

Figure 6-74   Commercial planning for Terminal 2 at Cangyuan Washan Airport

6.2.3.5 Green Technologies: Energy Saving for Ecological Airport

Under the low-latitude plateau subtropical monsoon climate, the architectural design features large eaves and shaded areas, combined with wind pressure analysis to optimize the design of facade openings, effectively organizing natural ventilation. The use of Low-E glass plus passive external shading eaves ensures the transparency of the terminal building while effectively reducing direct sunlight and preventing glare. The optimization and enhancement of the indoor lighting, combined with the design of skylights and glass curtain walls, achieve a comfortable indoor light environment with ample and even natural lighting, significantly reducing lighting energy consumption. The building's thermal bridges are minimized through structural optimization, enhancing insulation and thermal performance of the roof and exterior walls. (Fig. 6-75)

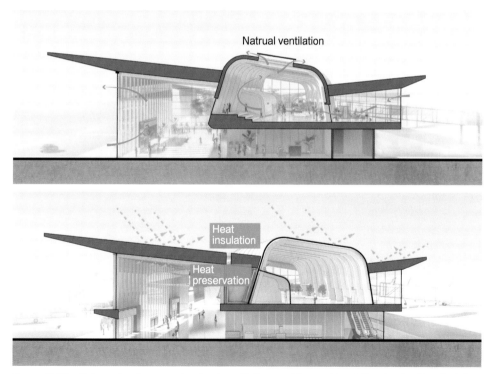

Figure 6-75 Schematic diagram of natural ventilation and thermal insulation for the Terminal 2 at Cangyuan Washan Airport

6.2.3.6 Unique Experience for Passengers: Cave Paradise in Wooden Drum, Secret Wa Homeland

The design of the Cangyuan Washan Airport Terminal 2 is a clear design manifesto for the fourth-generation regional airport terminals. It will create a new cultural calling card and open window for the Great Wa Mountain Area, allowing every visitor to Cangyuan to experience its unique charm, becoming a solid foundation for the economic takeoff of the region. The expanded and renovated Cangyuan Washan Airport will bring the world to the secret Wa homeland, and also bring the mysterious Washan to the world! (Figs. 6-76 ~ 6-78)

Figure 6-76   Perspective rendering of the facade of Terminal 2 at Cangyuan Washan Airport

Figure 6-77 Perspective rendering of the side facade of Terminal 2 at Cangyuan Washan Airport

Figure 6-78 Perspective rendering of the main facade of Terminal 2 at Cangyuan Washan Airport

6.2.4 Bengbu Airport Terminal: Clam and Pearl Nourished by Huaihe River

6.2.4.1 Overview: The Airport on the North-South Divide

Bengbu is also called "Pearl City," located in the northern plain of Anhui, stands on the dividing line between the south and north of China. The site of Bengbu Airport is located in Huaiyuan County, about 36 kilometers from the central area of Bengbu, 110 kilometers from Hefei Xinqiao Airport, and 193 kilometers from Nanjing Lukou International Airport. A composite transportation network that connects the north and south and radiates to the surrounding areas will be formed around Bengbu Airport, creating an integrated transportation system of air and rail.

In the future, Bengbu Airport will become not only the core of an efficient and convenient modern transportation system but also the first stop for people to understand Bengbu, carrying the mission of being the "Window of Pearl City, Impression of Bengbu."

6.2.4.2 Master Planning

Bengbu Airport is positioned as a 4C domestic regional airport in the near term and a 4E domestic medium-sized regional airport in the long term. The near-term construction scale of the terminal building is 30,000 square meters, which can serve an annual passenger throughput of 2.1 million person-times by 2040. The long-term target year is 2050, serving 3.5 million person-times passengers annually.

The near-term airfield level of Bengbu Airport is 4C, with a runway that is 2,600 meters long and 45 meters wide. In the long term, the runway will be extended to 3,200 meters to the west, and a parallel taxiway of the same length as the runway and 4 rapid exit taxiways will be set up. In the near term, there will be 15 apron parking stands and a 20,000-square meter parking lot. In the long term, the terminal building will be expanded to 50,000 square meters, the apron parking stands will be increased to 24, and the parking lot will be expanded to 37,000 square meters. (Figs. 6-79 ~ 6-82)

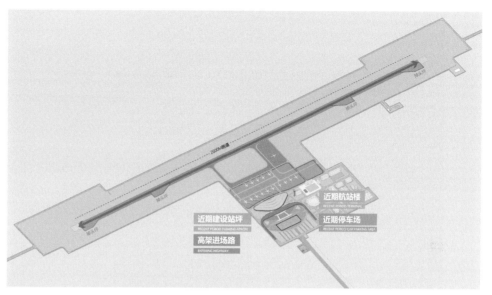

Figure 6-79　Near-term master planning of Bengbu Airport

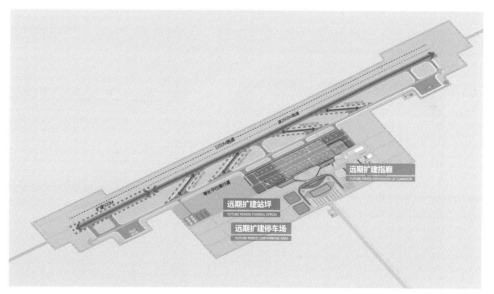

Figure 6-80　Long-term master planning of Bengbu Airport

Figure 6-81　Bird eye view of the landside of Bengbu Airport terminal in near term

Figure 6-82　Bird eye view of the landside of Bengbu Airport terminal in long term

6.2.4.3 Airside Planning

Taking into account factors such as air-land balance, the number of aircraft stands, land use intensity, and operational efficiency, the plan adopts a front-row configuration. This meets the near- and long-term development needs of Bengbu Airport, while achieving multiple goals such as the most convenience for passengers, operation and management.

A row of near parking stands facing the terminal building operates simply, without disturbing each other, and aircraft can quickly reach the near parking stands after leaving the runway. A row of remote parking stands is arranged behind the near stands, which can be used in coordination with the near stands to increase turnover rate and airbridge usage rate. The construction of Class E parking stands in the long term, with reasonable angle settings, not only saves the depth of the apron but also avoids the impact of Class E aircraft operations on the Class C apron. (Figs. 6-83, 6-84)

6.2.4.4 Landside Transportation Planning

Landside transportation planning of Bengbu Airport features three characteristics:

① Double-level one-way circulation system: The landside traffic in front of the terminal building is organized in a "counterclockwise one-way large circulation," with departure traffic on the upper level and arrival traffic on the lower level, avoiding cross confusion and ensuring orderly and efficient operations.

② Multi-curbside design: A total of five curbsides are set up in front of the terminal building. The elevated departure level is divided into inner and outer curbsides, with the inner curbside serving social

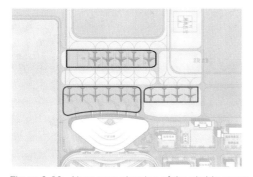

Figure 6-83   Near-term planning of the airside apron   Figure 6-84   Long-term planning of the airside apron

buses and special vehicles; the outer side is for social vehicles and taxi departure curbsides. The arrival level is a ground-level passenger pick-up system, also divided into inner and outer curbsides, with the inner downstream arranged for regular bus stops and the upstream arranged for taxi queuing points. One curbside is set up on the side of the parking lot near the terminal building, providing a space for social vehicles and online taxis to pick up and drop off passengers.

③ Priority is given to public transport and high-passenger-capacity vehicles, with buses stopping on the inner side and small vehicles organized on the outer side.

In terms of parking lot design, all social vehicles, online taxis, and rental vehicles are organized to pick up passengers within the parking lot. To be specific: ① Adopt a parking mode perpendicular to the terminal building, which is conducive to direct contact between passengers and the terminal building after parking. ② The parking lot clearly divides large, medium, and small vehicles, with short-term, long-term, online taxi, and rental car areas for zoned parking.

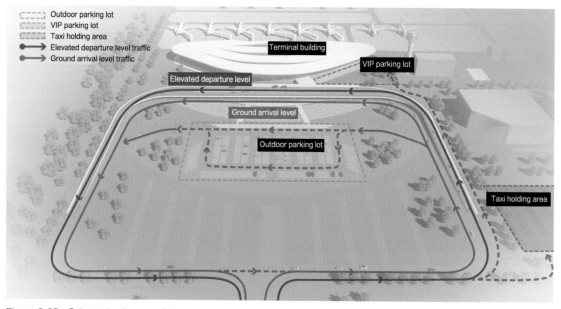

Figure 6-85　Schematic diagram of the double-level one-way large circulation system at landside

For Bengbu Airport, traffic flow lines were arranged according to vehicle types. ① Taxis: Drop off passengers at the outer curbside of the elevated departure level and leave, or enter the waiting area, and after dispatching, go to the taxi pick-up point on the arrival level to queue for passengers; ② Social vehicles and online taxis: Drop off passengers at the outer curbside of the elevated level or in the parking lot, and pick up passengers uniformly inside the parking lot and then leave; ③ Airport shuttle buses: Directly drive to the fixed position on the inner curbside in front of the first floor to pick up passengers and leave at fixed times; ④ VIP vehicles: Departing VIPs are taken to the VIP lounge via a dedicated lane, and arriving VIP vehicles join the unidirectional circulation system in front of the building to leave. (Figs. 6-85 ~ 6-94)

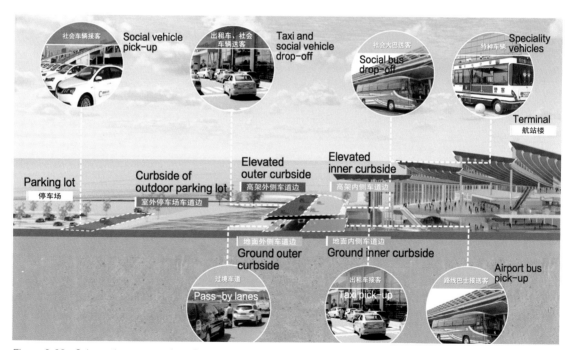

Figure 6-86　Schematic cross-section of the layout along the landside curbside

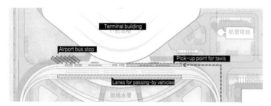

Figure 6-87　Plan of the curbside layout at the ground arrival level at landside

Figure 6-88　Plan of the curbside layout in the ground-level parking lot at landside

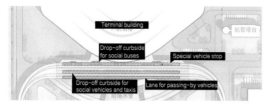

Figure 6-89　Plan of the curbside layout at the elevated departure level at landside

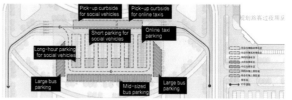

Figure 6-90　Plan of the layout of the ground-level parking lot at landside

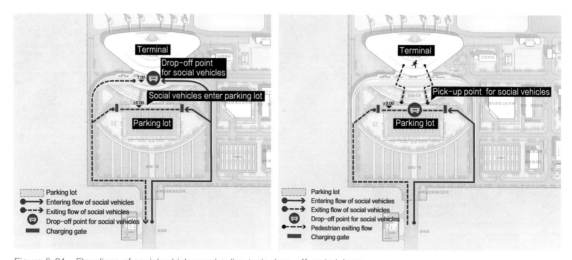

Figure 6-91　Flow lines of social vehicles and online taxis drop-off and pick-up

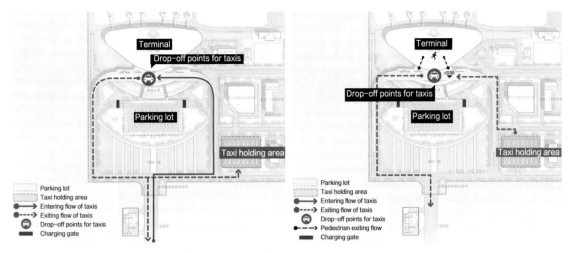

Figure 6-92 Flow lines for taxi drop-off, holding, and pick-up

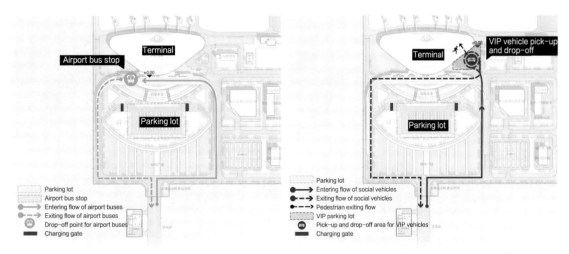

Figure 6-93 Flow lines for airport shuttle buses drop-off and pick-up

Figure 6-94 Flow lines for VIP vehicles drop-off and pick-up

6.2.4.5 Architectural Design Outlines

The design of terminal building started from two main points: a modern composite and efficient transportation hub, and a city window that showcases regional culture. The design team focused on culture, experience, efficiency, and ecology to build a fourth-generation regional airport terminal, adopting the following four major outlines:

1. Pearl in the Clam, Culture Gives Shape

Huaihe River and the pearl clam are not only cultural images with a strong sense of regional identity of Bengbu, but also symbols of Bengbu's long history and profound culture. (Fig. 6-95)

The design is based on the concept of "Clam and Pearl Nourished by Huaihe River," integrating the meaning of the Huaihe River's nourishment and Bengbu's pearls into a future pearl landmark in northern Anhui. The terminal building has a smooth and elegant curved form, with an orderly arrangement of roof arcs, unified in design with the water feature in front of the building, showcasing the shape of a clam by the water. (Figs. 6-96, 6-97) Additionally, an observation platform embedded with "pearl" spheres is added in the mezzanine of the terminal building, using the image of a pearl in the clam as the finishing touch, highlighting local cultural context with modern architectural language. This leaves an indelible first impression on passengers departing from the landside and arriving from the airside.

Figure 6-95 Design Inspiration from "Shell Holding a Pearl"

Figure 6-96　Bird eye view of the airside of Bengbu Airport terminal

Figure 6-97 Renderings of the "shell holding a pearl" design of Bengbu Airport terminal from multiple angles

Besides, the screen printing glass facade with the texture of clam pattern, visible as one approaches the terminal building, and the light ceiling with shell interior textures, overlapped and interlaced under the roof after entering the terminal, subtly display Bengbu's unique cultural image and historical mark throughout the passenger flow. This allows passengers to fully feel the unique charm of Bengbu. (Figs. 6-98 ~ 6-100)

2. Enter the Shell to Seek Pearls, Unique Experience

The design takes the concept of "entering the shell to seek pearls," and in the departure process, the flowing lines further refined from the curved roof create a spatial experience of a cave within the shell, guiding passengers from the landside departure hall to the observation platform located in the mezzanine. This platform can serve as an aviation education base for young people and increase non-aviation operation income. Upon entering the boarding lounge, the "pearl" space located at the center condense the display of Bengbu's historical culture and the national intangible cultural heritage—the Flower

Figure 6-98 Interior rendering of the departure hall of Bengbu Airport terminal

Figure 6-99　Interior rendering of the center of the departure hall of Bengbu Airport terminal

Figure 6-100　Interior rendering of the baggage claim area of Bengbu Airport terminal

Drum Lantern, and combining it with featured commerce, becoming a concentrated display window for Bengbu's regional culture and history. (Figs. 6-101 ~ 6-106)

Figure 6-101　Sectional diagram of Bengbu Airport terminal

Figure 6-102　Rendering of the outdoor observation platform on the landside of the terminal

Figure 6-103 Rendering of the "pearl" area before security check

Figure 6-104 Rendering of the "Pearl" in boarding lounge

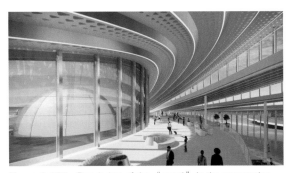

Figure 6-105 Rendering of the "pearl" in the mezzanine

Figure 6-106 Rendering of the "pearl" café interior

3. Efficient Zoning and Flexible Expansion

The terminal is designed as a two-level structure, with two main functional levels: The first level serves as the arrival area and VIP lounge, featuring baggage handling facilities, a baggage claim hall, and an arrival hall. The VIP lounge is located on the right side of the terminal, with a separate entrance and exit. The arrival hall provides access to the roadway and parking area in front of the terminal. The second level is accessed via elevated walkways and roadways, leading to the departure area. On both sides of the departure hall, check-in counters are set up, with a security check area in the middle. After security screening, passengers can proceed to the near boarding lounge or take escalators down to the first level's remote boarding lounge. There is also a partial mezzanine for a small airport museum and an outdoor landside observation deck, which can be accessed from the departure hall via escalators.

To accommodate passenger volume growth in the future, the terminal's facilities must have flexible expansion space. Based on the current facilities, provisions will be made for expanding check-in, security, and other services to meet a future annual throughput of 3.5 million person-times. (Figs. 6-107 ~ 6-109)

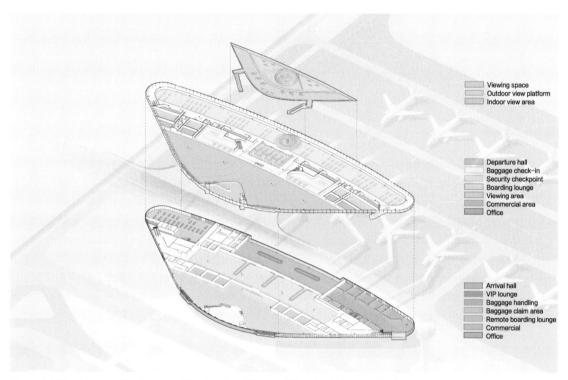

Figure 6-107　Axonometric diagram of the functional layout of Bengbu Airport terminal

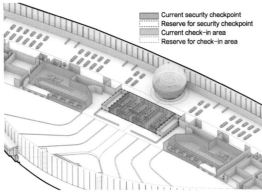 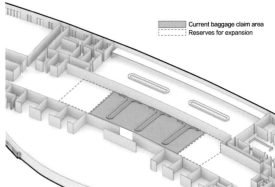

Figure 6-108　Flexible expansion for check-in and security checkpoint at Bengbu Airport terminal

Figure 6-109　Flexible expansion for baggage claim carousels at Bengbu Airport terminal

4. Energy Saving, Green Airport

Bengbu has distinct seasons with a significant monsoon climate and mild weather. The architectural design features a large overhanging roof that creates shaded areas. Coupled with wind pressure analysis, the roof design allows for the natural formation of operable high-side windows, effectively facilitating natural ventilation. The high-side window design of the terminal optimizes the indoor lighting environment, significantly reducing lighting energy consumption. Low-E glass combined with passive external shading from the overhanging roof ensures transparency of the terminal while effectively minimizing direct sunlight and glare. Structural optimization reduces thermal bridges in the building, enhancing the insulation performance of the roof and exterior walls. (Fig. 6-110)

5. Unique Experience for Passengers: "Clam and Pearl Nourished by Huaihe River"

At the beginning of the architectural design process, the team reached a consensus: only a clam that contains pearls is truly valuable. Therefore, the design concept is centered around "clam holding pearls." By ingeniously incorporating "pearl" spheres into the airside of the terminal and introducing an observation platform into the landside, the design aims to leave a lasting impression on passengers, with the expectation that Bengbu Airport will undoubtedly breathe new life into this city, and become a fresh window and calling card for this gem of the Huaihe River.

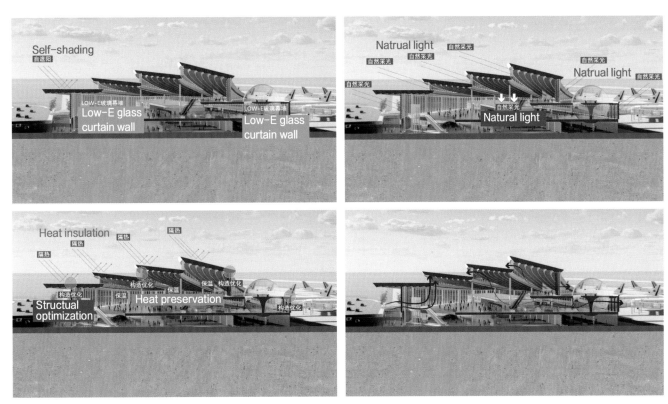

Figure 6-110　Schematic diagram of energy-saving analysis for Bengbu Airport terminal

SUMMARY

INNOVATIVE CONCEPTS AND EVALUATION OF
THE FOURTH-GENERATION (ERA OF EXPERIENCE)
REGIONAL AIRPORT TERMINAL

7.1 Concepts and Methods for Innovative Design of the Fourth-Generation Regional Airport Terminal

From the shifts observed in the four generational focuses of airport terminal design discussed in Chapter 6, it becomes evident that the design evolution of small- and medium-sized airports involves continuous innovation. This innovation aims to break through conventional thinking by building upon the foundations of previous exemplary design concepts, always closely aligned with the predominant concerns of each era. Therefore, the evolution in design for small- and medium-sized airport terminals does not outright dismiss the previous generational focuses, but rather represents an evolutionary process attempting to achieve breakthroughs and innovations from various perspectives.

Reflecting on the evolution of these four generational focuses in airport terminal design, the author identifies two essential elements that define the soul of terminal design in the era of experience:

① **Rooted in Regional and Cultural Contexts**

The emphasis on passenger experience in terminal design during the era of experience is not detached from its cultural roots. Instead, it closely aligns with the design focuses in era of culture, continuing to explore regional cultures and drawing nourishment from them. Deep and thoughtful excavation and interpretation of local cultural contexts should inform the design process. Moreover, in the era of experience for airport terminal design, there must be a deeper cultural interpretation than in previous eras, moving beyond mere abstraction and accumulation of cultural symbols. This approach should naturally translate into spatial atmospheres and cultural experiences that exhibit distinct local characteristics, serving as the wellspring of inspiration for terminal design in the era of experience.

② **Centered on Passenger Experience**

According to the "Action Outline for the Construction of 'Four Types of Airports' in China (2020—2035)" issued by the Civil Aviation Administration, future airport constructions will revolve around the principles of "safety, environmental friendliness, intelligence, and humanity," with a particular emphasis on humane considerations in the new era. Therefore, the ability to reflect human-centered care and to design with the traveler's experience at the forefront has become a new dimension of terminal building design in the era of experience, distinct from previous generations.

Currently, design methodologies focused on creating an unparalleled and comfortable passenger experience necessitate new requirements across various aspects of airport terminal design, such as form, space, structure, materials, and details. As experience constitutes a continuous human sensation, different facets of airport terminal design are no longer isolated entities, but are unified by the pursuit of enhancing passenger experience. This should serve as the core principle and guideline for the close coordination of all systems in the design of regional airport terminals.

Based on the essence of regional airport terminal design in the era of experience, and synthesizing with the experiences from the four cases of innovative terminal design discussed in Chapter 6 focusing on cultural experiences, the author proposes approaching design from five primary aspects: extraction and refinement of regional cultural elements, modern expression and adept application of cultural elements, creation of a distinctive and cohesive narrative experience, integration of indoor and outdoor design, and the organic fusion of experiential innovation with procedural workflows.

7.1.1 Extraction and Interpretation of Regional Culture

As previously mentioned, designing regional airport terminals in the era of experience requires a strong connection to regional and cultural contexts. The accuracy and skillful interpretation of regional cultural elements, and their ability to effectively reflect the essence of regional culture, are foundational to successful terminal designs in this era.

Through extensive analysis of exemplary cases and the author's team's practical experience in airport terminal design, the extraction of regional cultural elements can be approached from two perspectives: natural elements and humanistic connotations.

① Extraction of Regional Natural Elements

Natural features such as climate and topography often serve as significant regional characteristics, symbolizing and representing regional cultures. These elements are crucial in shaping regional identity, especially in areas with unique natural landscapes. For example, the snow-capped mountains of Xizang, the river valleys of Yunnan, and the deserts and mountains of Xinjiang each contribute a distinct cultural ambiance that defines the region. Therefore, in terminal design in these areas, extracting and refining natural elements serves as a pivotal starting point for interpreting regional culture.

② Extraction of Humanistic Connotations

Beyond natural elements, humanistic connotations are equally essential components of regional culture. Many small- and medium-sized airport terminals are located in regions rich in historical heritage or ethnic diversity. These places often express strong regional identities through cultural symbols, ethnic traditions, and historical narratives passed down over generations. Extracting these humanistic elements not only enriches the design process, but also plays a crucial role in shaping the terminal's identity and character.

7.1.2 Modern Expression and Skillful Application of Cultural Elements

After deeply excavating and extracting regional cultural elements, the second core challenge in terminal design is how to integrate these elements effectively to highlight the design. As discussed earlier, merely abstracting and stacking cultural symbols lacks sophistication, disconnects from architectural expression, and fails to resonate with travelers. In terminal design for the era of experience, it is crucial to abstract and translate cultural elements accurately and skillfully, expressing them through the architectural language to integrate cultural connotations into the spatial aspects of airport terminal design. This integration can generally be approached on three levels:

① External Form Reflecting Regional Culture

The external appearance of the terminal gives passengers the first impression of the overall space to shape their spatial experience initially. In the era of experience, the design should inherit and refine the abstraction of regional cultural symbols from the cultural era, combining them with modern principles of elegant and minimalist terminal architecture.

② Internal Space Reflecting Regional Culture

Currently, regional airport terminals often suffer from uniformity and excessive homogenization in interior design. This stems from a shallow interpretation and surface-level application of regional culture in previous terminal designs, resulting in a disconnection between external aesthetics and internal spatial functionality.

To address this, deeper excavation of regional cultures is needed to identity those that resonate with the terminal's internal space. This approach ensures the creation of a unique internal environment imbued with cultural significance, thereby avoiding uniformity issues.

③ Detailing Reflecting Regional Culture

Terminal detailing is crucial as it directly impacts passengers' physical interaction with space, contributing significantly to their overall experience. If external form and spatial layout represent macro- and meso-level strategies for translating regional cultural spaces and creating experiences, then detailed design with cultural elements provides a micro-level method to enhance these experiences.

In the era of experience, detailed cultural elements in terminal design should move away from mere stacking and collage seen in the era of culture. Instead, the focus should be on enhancing passenger experience by selecting culturally resonant elements that match the terminal's overall aesthetic and functional atmosphere. This approach ensures the creation of a culturally immersive experience at a scale that resonates closely with passengers in small- and medium-sized airport terminals.

7.1.3 Unique Experience for Passengers

Beyond shaping the external form of terminal buildings and translating cultural elements within internal spaces, creating experiences in small- and medium-sized airport terminals in the era of experience should transcend mere aggregation of architectural elements. Instead, it should treat passenger experience holistically, viewing it through a temporal and procedural lens. Starting with overall experience creation, this approach sequentially integrates external forms, internal spaces, structures, details, and other systems to craft a unified narrative spatial experience. The design practices outlined in Chapter 6 exemplify this approach.

7.1.4 Integrated Indoor and Outdoor Design

Designing spaces that evoke a unique cultural experience also imposes new requirements on the holistic design of airport terminals. Guided by a unified logic, the external form and spatial layout of the building should complement its internal spaces, fostering a progressive experiential relationship. An outstanding terminal design should convey a clear, coherent theme or evoke a distinct emotional response, ensuring strong spatial readability. This integrated approach to indoor and outdoor design represents a pivotal direction in terminal design for regional airport in the era of experience.

7.1.5 Integration of Space Innovation and Passenger Flow

Functional design not only underpins airport terminal buildings themselves, but also profoundly impacts travelers' immediate experiences, determining their comfort and sense of uniqueness within the terminal. In terminal design for the era of experience, innovation must aligns closely with functioanl flows; otherwise, it risks being superficial. Innovation should be rooted in regional culture and passenger experience, tightly integrated with passenger flows. This approach ensures efficient and convenient operation of terminal while offering travelers distinctive and innovative experiences.

7.2 Evaluation Aspects and Methods for Design of the Fourth-Generation Regional Airport Terminal

7.2.1 Evaluation Aspects

Over the generations, the focus of regional airport terminal building design has evolved, highlighting the importance of balancing functional rationality with cultural innovation. This section synthesizes the evaluation framework for terminal design in the era of experience, based on the core design principles and methods outlined earlier. This framework includes five key aspects: ① Refinement and capture of regional culture; ② Skillful expression of cultural elements; ③ Unique experience for passengers; ④ Integrated indoor and outdoor design; ⑤ Integration of space innovation and passenger flows.

Building upon successful design case studies, specific evaluation criteria for each aspect are articulated: ① The representativeness and communicative value of extracted regional culture; ② The ingenuity in applying regional cultural elements to the form and spatial composition of terminal building; ③ The ability of terminal's form and spatial composition to create distinctive experiential impact on passengers; ④ Consistency in presenting design concepts across interior and exterior of the terminal building; ⑤ The combination of experiential innovation with passenger flows, avoiding disjointed implementations.

Based on the evaluation criteria for the fourth generation of small- and medium-sized airport terminal designs as summarized above, a structured set of evaluation methods and tools can be developed. This systematic approach aims to provide comprehensive, scientific, and integrated support for evaluating future innovations in terminal design.

7.2.2 Evaluation Method: Quantification and Visualization

To quantify the evaluation criteria, a scoring system is proposed where each of the five aspects can earn a maximum of 20 points, totaling 100 points (Table 7-1). This approach allows for a comparative and visual assessment of the innovation levels across multiple terminal design projects. Additionally, visual tools such as radar charts can be employed to vividly depict the strengths and weaknesses of different design schemes based on the quantitative evaluations.

Table 7-1: Evaluation Form for Small- and Medium-Sized Airport Terminal Building Design

	Refinement and Capture of Regional Culture	Skillful Expression of Cultural Elements	Unique Experience for Passengers	Integrated Indoor and Outdoor Design	Integration of Space Innovation and Passenger Flows
Poor (0 ~ 5 points)					
Fair (6 ~ 10 points)					
Good (11 ~ 15 points)					
Excellent (16 ~ 20 points)					
Total score					

To effectively demonstrate how to use the evaluation method, the author conducted a simulated evaluation of the terminal design practice cases mentioned earlier, using a quantitative scoring table and a radar chart.

Zhanjiang Airport

Aspects	Refinement and Capture of Regional Culture	Skillful Expression of Cultural Elements	Unique Experience for Passengers	Integrated Indoor and Outdoor Design	Integration of Space Innovation and Passenger Flows
Description	The extraction of the "ray fish" element can well reflect the coastal culture of Zhanjiang	The combination of the "ray fish" element with the building shape is quite ingenious	The spatial experience within the terminal is good but lacks uniqueness	The external form and the internal space have been considered as a whole	A "light atrium" is innovatively introduced, combining well with the boarding process
Score	18	18	15	16	15
Total score	82				

Rizhao Shanzihe Airport

Aspects	Refinement and Capture of Regional Culture	Skillful Expression of Cultural Elements	Unique Experience for Passengers	Integrated Indoor and Outdoor Design	Integration of Space Innovation and Passenger Flows
Description	The extraction of the "shell" element can well reflect the coastal culture of Rizhao	The combination of the "shell" element with the building's curved shape is quite appropriate	The spatial experience within the terminal is comfortable but lacks uniqueness	The external form of "Shell of the Sea" and the indoor experience of "Rhythm of the Sea" share a strong sense of unity	The passenger flow is simple and reasonable but lacks innovation
Score	16	16	15	18	15
Total score	80				

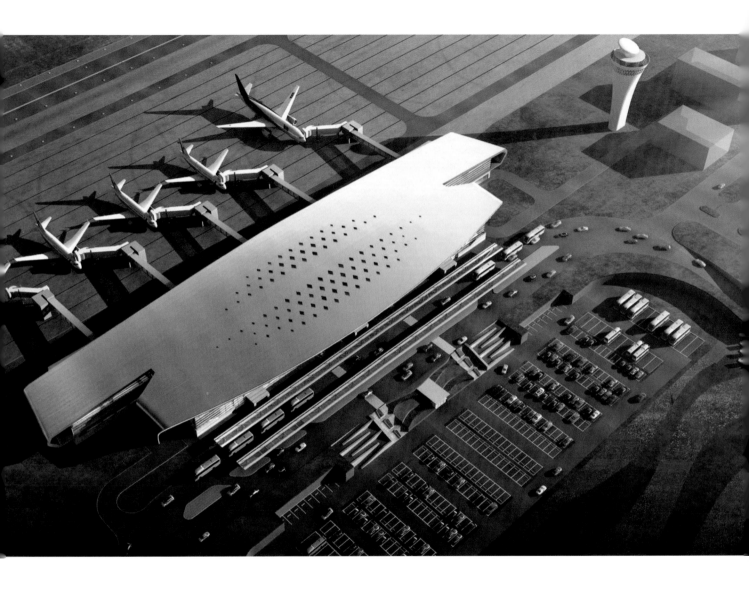

SUMMARY INNOVATIVE CONCEPTS AND EVALUATION OF THE FOURTH-GENERATION
(ERA OF EXPERIENCE) REGIONAL AIRPORT TERMINAL

Xizang Dingri Airport

Aspects	Refinement and Capture of Regional Culture	Skillful Expression of Cultural Elements	Unique Experience for Passengers	Integrated Indoor and Outdoor Design	Integration of Space Innovation and Passenger Flows
Description	The extraction of the "eagle of snowland" element is highly consistent with the site and local culture	Elements such as "eagle," "eagle feathers," and "khata" are ingeniously used in both the building shape and the internal space	The design provides a unique emotional experience of "looking up at the sacred mountain," envoking the excitement of "Climbing the Mount, Showcasing Passion"	The external shape concept of "eagle" and the internal space concept of "eagle feathers" are applied with a strong sense of unity	1) The triangular shape is highly compatible with the layout of arrival and departure halls as well as the vehicle lanes; 2) The use of "Rhata" elements is ingeniously combined with the sign system and passenger flows
Score	19	19	19	20	19
Total score	96				

Lancang Jingmai Airport

Aspects	Refinement and Capture of Regional Culture	Skillful Expression of Cultural Elements	Unique Experience for Passengers	Integrated Indoor and Outdoor Design	Integration of Space Innovation and Passenger Flows
Description	The extraction of the "green valley" and "gourd" elements is well aligned with local climate and the Wa culture	The use of "gourd" element to shape a unique boarding space is quite ingenious	The experience of "wandering in the sky courtyard to enjoy the pleasant climate" is relatively unique	The concept of "a gourd hiding in green valley" is expressed throughout from the external form to the internal space, showing good integrity	The sky courtyard is ingeniously combined with the boarding process to create a unique innovative experience
Score	18	18	18	18	18
Total score	90				

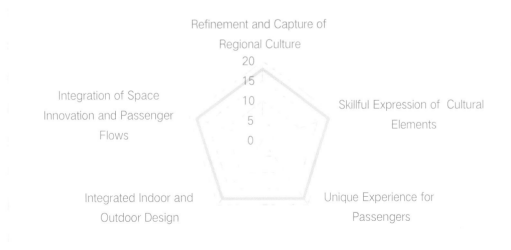

Lancang Jingmai Airport

Cangyuan Washan Airport (Proposal for Bidding)

Aspects	Refinement and Capture of Regional Culture	Skillful Expression of Cultural Elements	Unique Experience for Passengers	Integrated Indoor and Outdoor Design	Integration of Space Innovation and Passenger Flows
Description	The extraction of the "wooden drum" element and tale of "Sigangli" are highly aligned with Wa culture and resonate well	The spatialized use of "wooden drum" and "cave in Sigangli Tale" in the building's form is clever and fitting	The concepts of "Cave within Wooden Drum" and the "Airport Museum," derived from local culture, provide a unique experience	The external shape concept of "wooden drum" and the theme of "a paradise inside the drum" demonstrate unity	The innovative space of "Cave within Wooden Drum" serves as the boarding lounge, and the "Airport Museum" is well integrated with the waiting corridor
Score	19	19	20	18	19
Total score	95				

Dali Airport

Aspects	Refinement and Capture of Regional Culture	Skillful Expression of Cultural Elements	Unique Experience for Passengers	Integrated Indoor and Outdoor Design	Integration of Space Innovation and Passenger Flows
Description	1) The concept of "Bay of Wind" highly aligns with the featured local "Xiaguan Wind"; 2) The morphological feature of local traditional dwellings' roofs are skillfully extracted	1) The "wind" concept is expressed throughout the design of facade details quite ingeniously; 2) The sloped roof of local traditional dwellings is well expressed in the design of terminal building and transportation center	The introduction of green courtyards enhances the quality of the terminal space	The light and flowing shape with a transparent and simple internal space is unified in white color, demonstrating good integrity	The integration of landside view platforms and commercial spaces with the passenger flows shows a certain level of combination
Score	16	16	16	17	16
Total score	81				

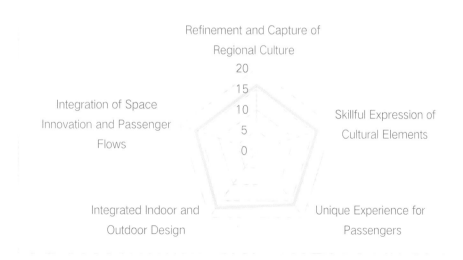

Dali Airport

SUMMARY INNOVATIVE CONCEPTS AND EVALUATION OF THE FOURTH-GENERATION
(ERA OF EXPERIENCE) REGIONAL AIRPORT TERMINAL

SUMMARY INNOVATIVE CONCEPTS AND EVALUATION OF THE FOURTH-GENERATION (ERA OF EXPERIENCE) REGIONAL AIRPORT TERMINAL

Bengbu Airport

Aspects	Refinement and Capture of Regional Culture	Skillful Expression of Cultural Elements	Unique Experience for Passengers	Integrated Indoor and Outdoor Design	Integration of Space Innovation and Passenger Flows
Description	The extraction of "shell" and "pearl" elements is highly consistent with the local culture of Bengbu	The external shape concept of "shell" and the spatial design focusing on "pearl" both show ingenious uses of cultural elements	The overall experience of "entering the shell to seek pearls" and the farewell interaction between landside and airside both have strong uniqueness	The external shape and the texture of the large indoor ceiling fully interpret the concept of "shell"	The landside view platform is connected with the airside commercial space of boarding lounge via the "pearl" spaces, which is closely integrated with the passenger flow
Score	18	18	18	17	20
Total score	91				

Illustration Sources

1 Introduction

Illustrations in this chapter are all drawn by the author.

2 Planning: General Composition and Basic Elements of Regional Airports

Figure 2-1: Cited from Harbin Taiping International Airport, Airfield Area Management Department, "Flight Area Knowledge"

Figure 2-3: Cited from "Civil Airport Airfield Area Technical Standards" (MH 5001—2021)

Figure 2-13: Cited from Paris Charles de Gaulle Airport Official Website

Figure 2-14: Cited from the official website of Daocheng Yading Airport, https://cn.yadingtour.com/wit/line

Figure 2-15: Cited from the official website of Beihai Fucheng Airport, https://cont.airport.gx.cn/index.php?m=mobile&siteid=5

Figure 2-23: Internet image, cited from https://www.cnipa.gov.cn/

The rest of illustrations in this chapter are drawn by the author.

3 Functions: Functional Composition and Design Outlines of Terminal Building

Figure 3-1: Original picture from "Civil Airport Terminal Building Design Fire Protection Code" (GB 1236—2017), redrawn by the author

Figure 3-29 left, Figure 3-54, Figure 3-63, Figures 3-68~3-70: Internet images, sources now unknown

Figure 3-40, Figure 3-41: Quoted from the security gate manufacturer's information

Figure 3-45: Original picture from the official website of AES Engineering, https://aesengr.com/lighting/lighting-case-studies/nanaimo-airport-terminal

Figure 3-46: Cited from the official website of Trans Asia Airport Lounge, https://www.plazapremiumlounge.com/zh-hk/find/europe/finland/helsinki/helsinki-airport/non-schengen-area- departures-terminal-two

Figure 3-47: Internet image, cited from https://www.traveller.com.au/brisbane-international-airport-terminal-to-get-45m-facelift-2yq04

The rest of illustrations in this chapter are drawn or photographed by the author.

4 Spatial Composition: Spatial Composition of One-and-a-Half-Story Terminal Building

Illustrations in this chapter are all drawn by the author.

5 Innovation: Design Innovation of Regional Airport Terminal Building

Figures 5-1 ~ 5-3: © b720 Fermín Vázquez Arquitectos

Figure 5-4, Figure 5-5: ©Román Viñoly Architects

Figure 5-6, Figure 5-7: ©andramatin

Figure 5-8, Figure 5-9, Figure 5-58: ©Group3Architects

Figure 5-10, Figure 5-11: ©FUKSAS

Figure 5-12, Figure 5-13: ©mad

Figure 5-14: Cited from Yidian Zixun, https://www.yidianzixun.com/article/0MdP7kgT?s=mochuang&appid=s3rd_mochuan&toolbar=&ad=&utk=

Figure 5-15: ©China Architecture Design & Research Group (CADG)

Figure 5-16, Figure 5-17: ©CITIC General Institute of Architectural Design & Research Co.Ltd. (Project Design: CITIC General Institute of Architectural Design & Research Co.Ltd.)

Figure 5-18, Figure 5-19, Figure 5-63, Figure 5-64: ©Integrated Design Associates

Figure 5-20, Figure 5-43: ©Nelson Kon

Figure 5-21, Figure 5-44: ©Biselli+Katchborian Arquitetos (Project Design: Biselli +Katchborian Arquitetos)

Figure 5-22: Internet image, source now unknown

Figure 5-23, Figure 5-24: Cited from the official website of Nagoya Chubu International Airport, https://www.centrair.jp/tzh/event/enjoy/flightpark/index.html

Figure 5-25: Internet image, source now unknown

Figure 5-26: ©HOK

Figures 5-27 ~ 5-29: ©ashleyhalliday

Figures 5-30 ~ 5-32: ©natkevicius

Figure 5-33, Figure 5-34: ©ZESO

Figure 5-35, Figure 5-36: ©3Dreid

Figure 5-37, Figure 5-38: ©MANN-SHINAR Architects

Figure 5-39, Figure 5-40: ©CCDI (Project Design: CCDI)

Figure 5-41, Figure 5-42: ©Ayala Arquitectos

Figure 5-45, Figure 5-46: ©Autoban

Figure 5-47: ©RSHP

Figure 5-48, Figure 5-49: ©Tuomas Uusheimo; Project Design: ALA Architects

Figure 5-50, Figure 5-51: ©Evoq Architecture

Figure 5-52, Figure 5-53: ©Nikken Sekkei LTD

Figure 5-54: Internet image, quoted from https://www.hippopx.com/zh/santorini-mediterranean-blue-greek-island-sea-travel-268353

Figure 5-55, Figure 5-56: ©AVW Architecture

Figure 5-57: Internet image, cited in "卡萨布兰卡哈桑二世清真寺游记", by: 全不能型tin, https://zhuanlan.zhihu.com/p/636839782

Figure 5-59, Figure 5-60: ©Studio of Pacific Architecture Limited

Figure 5-61, Figure 5-62: ©ductal

Figure 5-65, Figure 5-66: ©MGA | Michael Green Architecture (Project Design: MGA | Michael Green Architecture)

Figure 5-67: Drawn by the author

6 Tendency: Four Generations of Regional Airport Terminals and Innovative Practices of the Fourth Generation (Era of Experience)

Figure 6-1: Internet image, cited from https://pixabay.com/simon

Figures 6-5 ~ 6-7, Figure 6-15 top, Figure 6-19, Figure 6-25 left, Figure 6-38 left, Figure 6-39 left, Figure 6-49, Figure 6-60, Figure 6-61 left, Figure 6-62 left, Figure 6-95: Internet images, sources now unknown

The rest of illuatrtions in this chapter are photographed or drawn by the author.

7 Summary: Innovative Concepts and Evaluation of the Fourth-Generation (Era of Experience) Regional Airport Terminal

All illustrations in this chapter are drawn by the author.

Image Copyright Statement:

All images in this book are copyrighted by the original author, and all commercial use requires the consent of the original author or authorized website. In the process of preparing this book for public publication, the author of this book has been actively contacting the image copyright holders through different ways and has obtained the consent of some of them. For those who cannot be contacted, please send an email to mmh28775@ecadi.com, and we look forward to getting in touch with you. We would like to express our gratitude to all the licensed image copyright holders and related architectural firms.

Postscript

      This book is based on the outcomes of the project "Key Technology Research on the Design of One-and-a-Half-Story Terminal Building for Small Regional Airports" of Arcplus in 2023, and it is also a summary and condensation of the experience of the airport team of Southwest Division of ECADI, who have been engaged in a large number of airport design and engineering constructions in the past years.

      The first time the author put forward the concept of "the fourth-generation regional airport terminal" was in 2020 when preparing for Lancang Jingmai Airport bidding proposal. At that time, the author's team made an inventory of China's regional airports, and found that the construction of domestic regional airports had gone through many stages, with different requirements and focuses in each stage.

      The earliest requirements of regional airport terminals stayed at the process level, as long as the check-in, security, waiting, boarding and other passenger processes were included; the building itself was just a shell without high requirements for modeling. With the Reform and Opening Up, from the end of the 1990s, airports in China have been regarded as the aerial gateway of city, and have seen increasing requirements on their forms. There have been many tall and magnificent regional airport terminals, which are mostly large roofs and curved roofs. Due to the limited volume of those terminal buildings, most renovation was only based on the original small volume, leading to the general similarity of the overall modeling of the terminal building. After the 2010s, the terminal building has been given the mission of displaying regional culture, but the cultural expression of terminal in this period was mostly too figurative and straightforward modeling, feaured in symbols and superimposed collage, resulting in the disconnection of cultural expression of the exterior of the building with the interior space. The Dingri Airport terminal design in 2018 was a turning point, which was based on deep excavation of regional culture, and skillfully used it to realize an indoor and outdoor integrated design of local cultural elements; at the same time, the design studied Dingri Airport passenger profiles, and integrated them with the main process of terminal building to show local culture, clearly determining the design requirements for the "unique passenger experience."

With the establishment of the concept of "the fourth-generation regional airport terminal," the author's team has studied excellent cases of Chinese and foreign regional airports, and subsequently participated in the design practice of Cangyuan Washan Airport in Yunnan Province and Bengbu Airport in Anhui Province. In 2019, Arcplus' "Key Technology Research on the Design of One-and-a-Half-Story Terminal Building for Small Regional Airports" project was launched. By 2023, the project had summarized a set of innovative design concepts and evaluation methods for the "fourth-generation" regional airport terminals. In this process, the author's team invested and paid a lot, and also experienced the test of three years of COVID-19 epidemic. All these achievements would not be done without the hard work of all members of the airport design team of Southwest Division of ECADI.

The author would express special thanks to the following colleagues of ECADI Southwest Division: Deputy Chief Architect, Zhang Hongbo; Directors of Architecture, Tang Shuai and Yu Kai; Deputy Directors of Architecture, Li Senzi and Xu Wujian; Architects Xu Dan, Liao Wenyi, Zhao Zhuang, Peng Rui, Wang Qiao, Shan Zhizhong, Wang Yi, Li Yunfei, Li Zelin, Cai Weijie, Zou Yuhang, Lai Yizhen, Tan Yueqing, Chang Wenyu, and Li Tianci; Mechanical and Electrical Designers Ren Guojun, Liao Kaifeng, Chun Zongjing, and Xijunlin, etc. The author would also like to thank Ma Menghan for her efforts in communicating with the publisher during the book publishing process!

Thanks to everyone who contributed to the specialization of airports!